T0234094

DSP for
MATLAB™ and LabVIEW™
Volume IV: LMS Adaptive Filtering

SYNTHESIS LECTURES ON SIGNAL PROCESSING

Editor
José Moura, Carnegie Mellon University

DSP for MATLAB™ and LabVIEW™ Volume IV: LMS Adaptive Filtering
Forester W. Isen

ISBN: 978-3-031-01403-1   paperback
ISBN: 978-3-031-02531-0   ebook

DOI 10.1007/978-3-031-02531-0

A Publication in the Springer series
SYNTHESIS LECTURES ON SIGNAL PROCESSING

Lecture #7
Series Editor: José Moura, Carnegie Mellon University

Series ISSN
Synthesis Lectures on Signal Processing
Print 1932-1236   Electronic 1932-1694

DSP for MATLAB™ and LabVIEW™
Volume IV: LMS Adaptive Filtering

Forester W. Isen

SYNTHESIS LECTURES ON SIGNAL PROCESSING #7

ABSTRACT

This book is Volume IV of the series DSP for MATLAB™ and LabVIEW™. Volume IV is an introductory treatment of LMS Adaptive Filtering and applications, and covers cost functions, performance surfaces, coefficient perturbation to estimate the gradient, the LMS algorithm, response of the LMS algorithm to narrow-band signals, and various topologies such as ANC (Active Noise Cancelling) or system modeling, Noise Cancellation, Interference Cancellation, Echo Cancellation (with single- and dual-H topologies), and Inverse Filtering/Deconvolution. The entire series consists of four volumes that collectively cover basic digital signal processing in a practical and accessible manner, but which nonetheless include all essential foundation mathematics. As the series title implies, the scripts (of which there are more than 200) described in the text and supplied in code form (available via the internet at http://www.morganclaypool.com/page/isen) will run on both MATLAB and LabVIEW. The text for all volumes contains many examples, and many useful computational scripts, augmented by demonstration scripts and LabVIEW Virtual Instruments (VIs) that can be run to illustrate various signal processing concepts graphically on the user's computer screen. Volume I consists of four chapters that collectively set forth a brief overview of the field of digital signal processing, useful signals and concepts (including convolution, recursion, difference equations, LTI systems, etc), conversion from the continuous to discrete domain and back (i.e., analog-to-digital and digital-to-analog conversion), aliasing, the Nyquist rate, normalized frequency, sample rate conversion, and Mu-law compression, and signal processing principles including correlation, the correlation sequence, the Real DFT, correlation by convolution, matched filtering, simple FIR filters, and simple IIR filters. Chapter 4 of Volume I, in particular, provides an intuitive or "first principle" understanding of how digital filtering and frequency transforms work. Volume II provides detailed coverage of discrete frequency transforms, including a brief overview of common frequency transforms, both discrete and continuous, followed by detailed treatments of the Discrete Time Fourier Transform (DTFT), the z-Transform (including definition and properties, the inverse z-transform, frequency response via z-transform, and alternate filter realization topologies including Direct Form, Direct Form Transposed, Cascade Form, Parallel Form, and Lattice Form), and the Discrete FourierTransform (DFT) (including Discrete Fourier Series, the DFT-IDFT pair, DFT of common signals, bin width, sampling duration, and sample rate, the FFT, the Goertzel Algorithm, Linear, Periodic, and Circular convolution, DFT Leakage, and computation of the Inverse DFT). Volume III covers digital filter design, including the specific topics of FIR design via windowed-ideal-lowpass filter, FIR highpass, bandpass, and bandstop filter design from windowed-ideal lowpass filters, FIR design using the transition-band-optimized Frequency Sampling technique (implemented by Inverse-DFT or Cosine/Sine Summation Formulas), design of equiripple FIRs of all standard types including Hilbert Transformers and Differentiators via the Remez Exchange Algorithm, design of Butterworth, Chebyshev (Types I and II), and Elliptic analog prototype lowpass filters, conversion of analog lowpass prototype filters to highpass, bandpass, and bandstop filters, and conversion of analog filters to digital filters using the Impulse Invariance and

Bilinear Transform techniques. Certain filter topologies specific to FIRs are also discussed, as are two simple FIR types, the Comb and Moving Average filters.

KEYWORDS

Higher-Level Terms:

LMS Adaptive Filter, Least Mean Square, Active Noise Cancellation (ANC), Deconvolution, Equalization, Inverse Filtering, Interference Cancellation, Echo Cancellation, Dereverberation, Adaptive Line Enhancer (ALE)

Lower-Level Terms:

Single-H, Dual-H, Gradient, Cost Function, Performance Surface, Coefficient Perturbation

This volume is dedicated to

The following humane organizations, and all who work for them and support them:

Rikki's Refuge (www.rikkisrefuge.org), Orange, Virginia
The Humane Society of Fairfax County (www.hsfc.org), Fairfax, Virginia
The Washington Animal Rescue League (WARL) (www.warl.org), Washington, DC
Angels for Animals (www.angelsforanimals.org), Youngstown, Ohio
and all similar organizations and persons

and to the feline companions who so carefully watched over the progress of the the entire series during its evolution over the past eight years:

Richard (1985-2001)
Blackwythe (1989-2001)
Tiger (1995? -)
Mystique (2000 -)
Scampy (2002 -)
Nudgy (2003 -)
Percy (2004 -)

Contents

Preface to Volume IV

0.1 INTRODUCTION

The present volume is Volume IV of the series DSP for *MATLAB*™ *and LabVIEW*™. The entire series consists of four volumes which collectively form a work of twelve chapters that cover basic digital signal processing in a practical and accessible manner, but which nonetheless include essential foundation mathematics. The text is well-illustrated with examples involving practical computation using m-code or MathScript (as m-code is usually referred to in LabVIEW-based literature), and LabVIEW VIs. There is also an ample supply of exercises, which consist of a mixture of paper-and-pencil exercises for simple computations, and script-writing projects having various levels of difficulty, from simple, requiring perhaps ten minutes to accomplish, to challenging, requiring several hours to accomplish. As the series title implies, the scripts given in the text and supplied in code form (available via the internet at `http://www.morganclaypool.com/page/isen`) are suitable for use with both MATLAB (a product of The Mathworks, Inc.), and LabVIEW (a product of National Instruments, Inc.). Appendix A in each volume of the series describes the naming convention for the software written for the book as well as basics for using the software with MATLAB and LabVIEW.

0.2 THE FOUR VOLUMES OF THE SERIES

The present volume, Volume IV of the series, LMS Adaptive Filtering, begins by explaining cost functions and performance surfaces, followed by the use of gradient search techniques using coefficient perturbation, finally reaching the elegant and computationally efficient Least Mean Square (LMS) coefficient update algorithm. The issues of stability, convergence speed, and narrow-bandwidth signals are covered in a practical manner, with many illustrative scripts. In the second chapter of the volume, use of LMS adaptive filtering in various filtering applications and topologies is explored, including Active Noise Cancellation (ANC), system or plant modeling, periodic component elimination, Adaptive Line Enhancement (ADE), interference cancellation, echo cancellation, and equalization/deconvolution.

Volume I of the series, Fundamentals of Discrete Signal Processing, consists of four chapters. The first chapter gives a brief overview of the field of digital signal processing. This is followed by a chapter detailing many useful signals and concepts, including convolution, recursion, difference equations, etc. The third chapter covers conversion from the continuous to discrete domain and back (i.e., analog-to-digital and digital-to-analog conversion), aliasing, the Nyquist rate, normalized frequency, conversion from one sample rate to another, waveform generation at various sample rates from stored wave data, and Mu-law compression. The fourth and final chapter of Volume I introduces the reader to many important principles of signal processing, including correlation, the correlation sequence, the Real DFT, correlation by convolution, matched filtering, simple FIR filters, and simple IIR filters.

Volume II of the series is devoted to discrete frequency transforms. It begins with an overview of a number of well-known continuous domain and discrete domain transforms, and covers the DTFT (Discrete Time Fourier Transform), the DFT (Discrete Fourier Transform), Fast Fourier Transform (FFT), and the z-Transform in detail. Filter realizations (or topologies) are also covered, including Direct, Cascade, Parallel, and Lattice forms.

Volume III of the series is devoted to Digital Filter Design. It covers FIR and IIR design, including general principles of FIR design, the effects of windowing and filter length, characteristics of four types of linear phase FIR, Comb and MA filters, Windowed Ideal Lowpass filter design, Frequency Sampling design with optimized transition band coefficients, Equiripple FIR design, and Classical IIR design.

0.3   ORIGIN AND EVOLUTION OF THE SERIES

The manuscript from which the present series of four books has been made began with an idea to provide a basic course for intellectual property specialists and engineers that would provide more explanation and illustration of the subject matter than that found in conventional academic books. The idea to provide an accessible basic course in digital signal processing began in the mid-to-late 1990's when I was introduced to MATLAB by Dan Hunter, whose graduate school days occurred after the advent of both MATLAB and LabVIEW (mine did not). About the time I was seriously exploring the use of MATLAB to update my own knowledge of signal processing, Dr. Jeffrey Gluck began giving an in-house course at the agency on the topics of convolutional coding, trellis coding, etc., thus inspiring me to do likewise in the basics of DSP, a topic more in-tune to the needs of the unit I was supervising at the time. Two short courses were taught at the agency in 1999 and 2000 by myself and several others, including Dr. Hal Zintel, David Knepper, and Dr. Pinchus Laufer. In these courses we stressed audio and speech topics in addition to basic signal processing concepts. Thanks to The Mathworks, Inc., we were able to teach the in-house course with MATLAB on individual computers, and thanks to Jim Dwyer at the agency, we were able to acquire several server-based concurrent-usage MATLAB licenses, permitting anyone at the agency to have access to MATLAB. Some time after this, I decided to develop a complete course in book form, the previous courses having consisted of an ad hoc pastiche of topics presented in summary form on slides, augmented with visual presentations generated by custom-written scripts for MATLAB. An early draft of the book was kindly reviewed by Motorola Patent Attorney Sylvia Y. Chen, which encouraged me to contact Tom Robbins at Prentice-Hall concerning possible publication. By 2005, Tom was involved in starting a publishing operation at National Instruments, Inc., and introduced me to LabVIEW with the idea of possibly crafting a book on DSP to be compatible with LabVIEW. After review of a manuscript draft by a panel of three in early 2006, it was suggested that all essential foundation mathematics be included so the book would have both academic and professional appeal. Fortunately, I had long since retired from the agency and was able to devote the considerable amount of time needed for such a project. The result is a book suitable for use in both academic and professional settings, as it includes essential mathematical formulas and concepts as well as simple or "first principle" explanations that help give the reader a gentler entry into the more conventional mathematical treatment.

This double-pronged approach to the subject matter has, of course, resulted in a book of considerable length. Accordingly, it has been broken into four modules or volumes (described above) that together form a comprehensive course, but which may be used individually by readers who are not in need of a complete course.

Many thanks go not only to all those mentioned above, but to Joel Claypool of Morgan&Claypool, Dr. C.L. Tondo and his troops, and, no doubt, many others behind the scenes whose names I have never heard, for making possible the publication of this series of books.

Forester W. Isen
January 2009

CHAPTER 1

Introduction To LMS Adaptive Filtering

1.1 OVERVIEW

1.1.1 IN PREVIOUS VOLUMES

The previous volumes of the series are Volume I, *Fundamentals of Discrete Signal Processing*, Volume II, *Discrete Frequency Transforms*, and Volume III, *Digital Filter Design*. Volume I covers DSP fundamentals such as basic signals and LTI systems, difference equations, sampling, the Nyquist rate, normalized frequency, correlation, convolution, the real DFT, matched filtering, and basic IIR and FIR filters. Volume II covers the important discrete frequency transforms, namely, the Discrete Time Fourier Transform (DTFT), the Discrete Fourier Transform (DFT), and the z-transform. Volume III considers the design of both FIR and IIR filters that are intended to have fixed characteristics for use in a static environment.

1.1.2 IN THIS VOLUME

There are many situations in which needed filter characteristics change rapidly, or in which the needed characteristics are not precisely known in advance. Adaptive filtering fills these needs by automatically changing filter coefficients to achieve some particular goal. In this volume, we take up the study of filters that can adapt their characteristics according to some criteria. In particular, our study will center on LMS Adaptive filtering, which is popular because of its low computational overhead. It has found applications in Active Noise Cancellation (ANC), echo cancellation in telephone systems, beamforming, narrowband signal attenuation or enhancement, equalizers, etc. In this volume, consisting of two chapters, we first examine several fundamental ideas which culminate in derivation of the LMS Algorithm. In the second chapter, we examine a number of common filter topologies or applications of LMS Adaptive filtering.

1.1.3 IN THIS CHAPTER

We begin our study of adaptive filtering by first investigating cost functions, performance surfaces, and the gradient, using a simple technique called **Coefficient Perturbation**. We'll consider several examples, including one- and two- independent-variable quadratic cost functions, including the problem of fitting a line to a set of points in a plane. The simplicity of coefficient perturbation will enable the reader to readily understand fundamental concepts and issues relevant to gradient-search-based adaptive filtering. We then derive a more efficient way of estimating the gradient, the simple, elegant, and efficient algorithm known as the **LMS (Least Mean Squared)** algorithm, which, with its various derivative algorithms, accounts for the majority of adaptive algorithms in commercial use. We'll derive the algorithm for a length-2 FIR, which readily generalizes to a length-N FIR. We then explore one of the weaknesses of the LMS algorithm, its need for a wide-bandwidth signal for proper algorithm function in certain topologies, such as ANC.

By the end of the chapter, the reader will have gained fundamental knowledge of the LMS algorithm and should be able (and, in the exercises at the end of the chapter, is expected) to write scripts to implement simple LMS adaptive FIR filters. The reader will then be prepared for the next chapter in which we explore the LMS algorithm in various useful applications, such as active noise cancellation, signal enhancement, echo cancellation, etc.

1.2   SOFTWARE FOR USE WITH THIS BOOK

The software files needed for use with this book (consisting of m-code (.m) files, VI files (.vi), and related support files) are available for download from the following website:

http://www.morganclaypool.com/page/isen

The entire software package should be stored in a single folder on the user's computer, and the full file name of the folder must be placed on the MATLAB or LabVIEW search path in accordance with the instructions provided by the respective software vendor (in case you have encountered this notice before, which is repeated for convenience in each chapter of the book, the software download only needs to be done once, as files for the entire series of four volumes are all contained in the one downloadable folder). See Appendix A for more information.

1.3   COST FUNCTION

A common problem is the need to fit a line to a set of test points. The test points, for example, might be data taken from an experiment or the like. We'll start by investigating how a "best fit" curve (or, in the example at hand, straight line) can be determined mathematically, i.e., what mathematical criterion can be used for an algorithm to determine when a candidate curve or line is better than another?

In Fig. 1.1, five test points (circles) are to have a horizontal line fitted in the "best" manner possible using purely mathematical criteria. For purposes of simplicity, the line is allowed to vary only in the vertical direction. In a real problem, the points would likely be irregularly placed and the line would have to assume different slopes as well as different vertical positions (we will examine line-modeling in just this way later in the chapter).

In Fig. 1.1, if we define the vertical location of the line as y; the distance from each upper point to the line is $(2 - y)$, and the distance from each of the two lower points to the line is y. We'll construct three alternative functions of y and compare their performance in determining the best value of y to place the line in an intuitively, satisfying location.

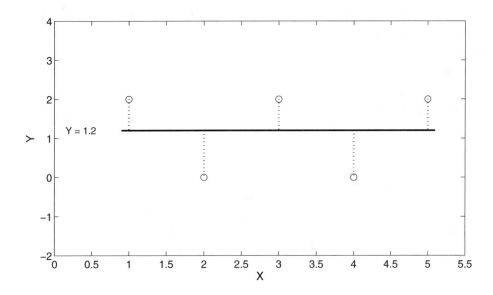

Figure 1.1: Five points in a plane to be modeled by a line $y = a$, where a is a constant.

For the first function, we'll use a first order statistic, that is to say, we'll sum the various distances and normalize by the number of distances (i.e., divide the sum by the number of distances), in accordance with Eq. (1.1):

$$F_1(y) = \frac{1}{5}(3(2 - y) + 2y) \tag{1.1}$$

For the second function, we'll also use a first order statistic, but we'll sum the magnitudes of the various distances and normalize. This results in Eq. (1.2):

$$F_2(y) = \frac{1}{5}(3(|2 - y|) + 2|y|) \tag{1.2}$$

For the third function, we'll use second-order statistics, i.e., the square of distances from the line to the various points to be modeled. To do this, we square each distance mentioned above, sum all squared distances, and normalize. We'll call this the MSE (mean squared error) method.

Using the MSE method, the function of distance looks like this:

$$F_3(y) = \frac{1}{5}(3(2 - y)^2 + 2y^2) \tag{1.3}$$

We can evaluate Eqs. (1.1) through (1.3) with a test range for y to see how well each function generates an intuitively satisfying location for the line between the two rows of test points. Letting the three functions given above be represented respectively by $F1$, $F2$, and $F3$, the results from running the following m-code are shown in Fig. 1.2:

```
y = -1:0.05:3;
F1 = (1/5)*(3*(2-y) + 2*(y));
F2 = (1/5)*(3*abs(2-y) + 2*abs(y));
F3 = (1/5)*(3*(2-y).^2 + 2*(y.^2));
```

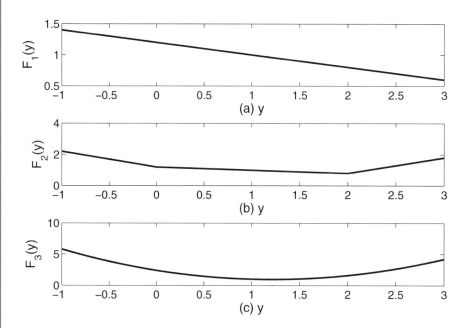

Figure 1.2: (a) The function $F_1(y)$ (see text) evaluated over the range y = -1:0.05:3; (b) Same, for function $F_2(y)$; (c) Same, for function $F_3(y)$.

We can refer to F_1, F_2, and F_3 as **Cost Functions** since there is a "cost" or value associated with placing the line at certain positions. As in purchases using money, the goal is to minimize the cost.

Intuitively, the best location for the line is somewhere between the three upper points and the two lower points, and probably a little closer to the three upper points.

Neither first order cost function (plots (a) and (b) of Fig. 1.2) does well. The first function does not provide any useful guidance; the second one does give guidance, but its minimum occurs when the line passes through the three upper points (when y = 2). This does not suit our intuitive notion of best placement.

The second order, or MSE cost function (plot (c) of Fig. 1.2) shows its minimum cost at y = 1.2, which seems a reasonable placement of the line. Notice also that there is a single, distinct minimum for the second order cost function. It is clear that the best cost function for this simple example is MSE or second order. This is true in general for many problems, and it is the cost function that we will be using in this and the following chapter in the development and use of various adaptive algorithms.

- Second order cost functions (or MSE functions), have single ("global") minima that produce an intuitively satisfying "best" result for many line and curve fitting problems.

1.4   PERFORMANCE SURFACE

Any of plots (a)-(c) of Fig. 1.2 may be described as **Performance Curves**. Each is a graph of an independent variable versus the corresponding value of a cost function. In situations where there are two or more independent variables, the cost function, graphed as a function of its independent variables, takes on three or more dimensions and hence is a surface rather than a simple curve. Hence, it is called a **Performance Surface**. We will see examples of the performance surface a little later, but first, we'll take a look at a couple of general methods for "reaching the bottom" of a performance curve algorithmically, that it to say, using a computer program.

For this kind of problem, in general, we will have one or more independent variables which when varied will cause the cost function to vary, and our goal is for a computer program or algorithm to change the values of the one or more independent variables in such a manner that eventually the cost function is minimized.

1.5   COEFFICIENT PERTURBATION

An intuitive way of determining how to adjust each independent variable to minimize the cost function is to let all of the independent variables assume certain fixed values (i.e., assign values to the variables) and then evaluate the cost function. Next, change one of the variables by a very small amount (this is called perturbation) while holding all the other variable values constant, and then evaluate the cost function. If the change to the particular variable in question was in the positive direction, for example, and the cost function increased in value, then logically, changing the value of that variable in the negative direction would tend to decrease the value of the cost function.

A mathematical way of describing this, for a cost function that depends on the single independent variable x, would be

$$CF = \frac{\Delta(Cost\,Fcn)}{\Delta x} = \frac{Cost\,Fcn(x + \Delta x) - Cost\,Fcn(x)}{\Delta x} \tag{1.4}$$

The ratio in Eq. (1.4) tells how much and in what direction the cost function varies for a unit change in the independent variable x. The limit of the ratio in Eq. (1.4) as Δx goes to zero is the derivative of the cost function CF with respect to x:

$$CF'(x) = \frac{d(CF)}{dx} = \lim_{\Delta x \to 0} \left(\frac{CF(x + \Delta x) - CF(x)}{\Delta x} \right)$$

This concept is easily extended to a cost function that is dependent upon two or more independent variables. Then if

$$y = f(x_1, x_2, ...x_N)$$

the partial derivative of y with respect to a particular independent variable x_i is

$$\frac{\partial(y)}{\partial(x_i)} = \lim_{\Delta x_i \to 0} \frac{f(x_1, x_2, ...(x_i + \Delta x_i), ...x_N) - f(x_1, x_2, ...x_N)}{\Delta x_i}$$

In computer work, there is a limit to precision, so making Δx_i smaller and smaller does not, beyond a certain point, improve the estimate of the partial derivative. Thus it is not possible (referring to the single-independent-variable example at hand) to obtain

$$\lim_{\Delta x \to 0} \frac{CF(x + \Delta x) - CF(x)}{\Delta x} \tag{1.5}$$

since, due to roundoff error, the true limit will in general not be monotonically approached or achieved. For most normal purposes, however, this is not a serious problem; a useful estimate of the partial derivative can be obtained. One method that can increase the accuracy from the very first estimate is to straddle the value of x. This is accomplished by dividing Δx by two and distributing the two parts about x as follows:

$$CF'(x) \simeq \frac{CF(x + \Delta x/2) - CF(x - \Delta x/2)}{\Delta x} \tag{1.6}$$

In Eq. (1.6), we have adopted the symbol $\simeq$ to represent the estimate or approximation of the partial derivative of $CF(x)$ with respect to x. Note that the two values of x at which the cost function is computed surround the actual value of x, and provide a better estimate of the slope of the cost function at x than would be provided by evaluating the cost function at x and $x + \Delta x$. This method is referred to as taking **Central Differences**, as described in Chapter 5 of Reference [1].

The **Gradient** of a function (such as our cost function CF) for N independent variables x_i is denoted by the **del** (∇) operator and is defined as

$$\nabla(CF) = \sum_{i=1}^{N} (\partial(CF)/\partial(x_i))\hat{u}_i \tag{1.7}$$

where $\hat{u}_i$ is a unit vector along a dimension corresponding to the independent variable x_i. The purpose of the unit vectors is to maintain each partial derivative as a distinct quantity. In the programs and examples below involving cost functions dependent on two or more independent variables, each partial derivative is simply maintained as a distinct variable.

Once all the partial derivatives have been estimated using perturbation, it is possible to update the current best estimate for each of the independent variables x_i using the following update equation:

$$x_i[k+1] = x_i[k] - \mu CF'(x_i, k) \tag{1.8}$$

where $CF'(x_i, k)$ means the estimate of the partial derivative of CF with respect to independent variable x_i at sample time k. Equation (1.8) may be described verbally thus: the estimate of the variable x_i at iteration $(k+1)$ is its estimate at iteration k minus the product of the partial derivative of the cost function with respect to x_i at iteration k and a small value μ, which helps regulate the overall size of the update term, $\mu CF'(x_i, k)$.

Too large a value of μ will cause the algorithm to become unstable and diverge (i.e., move away from the minimum of the cost function) while too small a value will result in protracted times for convergence. New estimates of the coefficients are computed at each sample time and hopefully the coefficient estimates will converge to the ideal values, which are those that minimize the cost function. Convergence to the true minimum of the performance surface, however, can only be reliably achieved if the performance surface is unimodal, i.e., has one global minimum. Some performance surfaces may possess local minima to which the gradient may lead, and the algorithm can fail to converge to the true minimum of the cost function. Fortunately, the problems we'll investigate do have unimodal performance surfaces. As we work through various examples, we'll see how various parameters and situations affect algorithm convergence.

1.6   METHOD OF STEEPEST DESCENT

Using the negative of the gradient to move down the performance surface to the minimum value of the cost function is called the **Method of Steepest Descent;** this is because the gradient itself points in the direction of steepest *ascent*, while the negative of the gradient, which we use in coefficient update Eq. (1.8) moves in precisely the opposite direction–the direction with the steepest slope down the performance surface.

The following discussion uses a simple quadratic cost function to explore a number of important concepts involved in adaptive processes, such as gradient (slope), step size, convergence speed, accuracy of gradient estimate, etc. In the several plots of Fig. 1.3, the cost function $CF(x)$ is

$$CF(x) = x^2 + 1$$

and the ultimate goal is to determine the value of x at which the $CF(x)$ is minimized, i.e., reduced to 1.0.

Plot (a) of Fig. 1.3 shows the first step in navigating a Performance Surface to its minimum using the method of steepest descent, in which an initial value of x was chosen arbitrarily to be 4.0. From this, the value of the gradient must be estimated and the estimate used to compute a new value of x which is (hopefully) closer to the value of x which minimizes the cost function.

In any of the plots of Fig. 1.3, values of the cost function versus x are plotted as small circles; the method of steepest descent may be understood by imagining the circles as being marbles, and the performance curve as a surface down which the marbles can roll under the pull of gravity. In

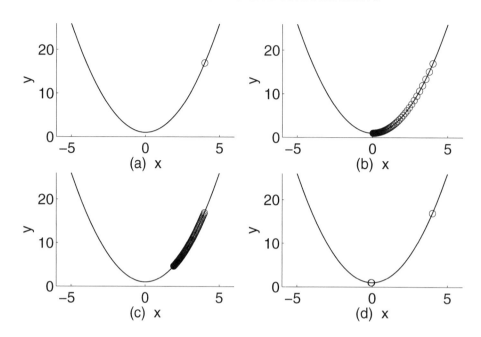

Figure 1.3: (a) Graph of cost function and first value of parameter x and the corresponding value of the cost function; (b) Successive estimated values of x and respective values of the cost function for 75 iterations; (c) Seventy-five iterations, attempting to reach the minimum value of the cost function using too small a value of StepSize (μ); (d) Three iterations using a weighting function for μ.

a more complex problem with the performance surface being a (three-dimensional) paraboloid-of-revolution rather than a simple two-dimensional curve, such a "marble" would roll straight down along the inside of the paraboloid (a bowl-shaped surface) toward the bottom.

The script

$$LVxGradientViaCP(FirstValX, StepSize,...$$

$$NoiseAmp, UseWgtingFcn, deltaX, NoIts)$$

(see exercises below) allows experimentation with the method of steepest descent using a second-order cost function having a single independent variable. A typical call would be:

LVxGradientViaCP(4,0.05,0.01,0,0.0001,50)

which results in Fig. 1.3, plot (a), for Iteration 1; the script is designed so that successive iterations of the algorithm are performed by pressing any key.

The first argument, *FirstValX*, is the starting value (the initial guess) for the only independent variable, x. *StepSize* is equivalent to μ, the parameter that regulates the size of the update term. *NoiseAmp* is the standard deviation of white noise that is added to the current estimate of x to more realistically simulate a real (i.e., noisy) situation (set it to zero if desired). The fifth argument, *deltaX* (Δx in the pseudocode below), is the amount by which to perturb the coefficients when estimating the gradient. The sixth argument, *NoIts*, is the number of iterations to do in seeking the value of x that minimizes the cost function. The variable *UseWgtingFcn* (UWF in the pseudocode below) will be discussed extensively below.

The heart of the algorithm, in pseudocode terms, is

$for\ \ n = 1 : 1 : NoIts$
$CF[n] = (x[n] - \Delta x/2)^2 + 1$
$TstCF = (x[n] + \Delta x/2)^2 + 1$
$\partial(CF)/\partial(x[n]) = (TstCF - CF[n])/\Delta x$
$if\ UWF = 1$
$\ \ x[n + 1] = x[n] - \mu CF[n](\partial(CF)/\partial(x[n]))$
$else$
$\ \ x[n + 1] = x[n] - \mu(\partial(CF)/\partial(x[n]))$
end
end

The pseudocode above parallels the preceding discussion exactly, i.e., implementing an algorithm that performs gradient estimation via central-differences-based coefficient perturbation. The product of $\partial(CF)/\partial(x[n])$ and μ (*StepSize*) is the amount by which the value of x is to be changed for each iteration. This quantity may be further modified (as shown above) by multiplying it by the current value of *CostFcn* ($CF[n]$) in order to cause the entire term to be large when the error (and hence *CF*) is large, and small when *CF* becomes small. This helps speed convergence by allowing the algorithm to take larger steps when it is far from convergence and smaller steps when it is close to convergence. The fourth input argument to the function call, *UseWeightingFcn* (UWF in the pseudocode above), determines whether or not the update term uses $CF[n]$ as a weighting function or not.

The coefficient update term, the product of $\partial(CF)/\partial(x[n])$, μ, and $CF[n]$ if applied, has great significance for convergence and stability. If the update term is too large, the algorithm diverges; if it is too small, convergence speed is too slow. In the following examples, we'll observe convergence rate and stability with the update term computed with and without the weighting function.

Example 1.1.   Use the script *LVxGradientViaCP* to demonstrate stable convergence.

We call the script with no weighting function and *StepSize* = 0.03:

LVxGradientViaCP(4,0.03,0.01,0,0.0001,75)

Figure 1.3, plot (b), shows the result. Convergence proceeds in a slow, stable manner.

- A useful line of experimentation is to hold all input arguments except *StepSize* (μ) constant, and to gradually increase μ while noticing how the algorithm behaves. For this set of input parameters (i.e., with no weighting function), the reader should note that the value of μ at which the algorithm diverges is about 1.0. If using a weighting function, the algorithm will behave much differently for a given value of μ. The following examples point out different significant aspects of algorithm performance based (chiefly) on the parameters of μ and *UseWgtingFcn*.

Example 1.2.   Demonstrate inadequate convergence rate due to too small a value of *StepSize*.

The result of the use of the same call as for the preceding example, but with μ = 0.005, is shown in Fig. 1.3, plot (c), in which 75 iterations are insufficient to reach the minimum value of the cost function.

Example 1.3.   Demonstrate stable overshoot by use of too large a value of *StepSize*, such as 0.9, with no weighting function.

An appropriate call is

LVxGradientViaCP(4,0.9,0.01,0,0.0001,50)

which results in overshooting the minimum, but eventually descending to it. Figure 1.4 shows what happens–the successive estimates of x oscillate around the ideal value of x, but the amplitude of oscillation gradually decays, leading to convergence to the minimum of the cost function.

Example 1.4.   Experimentally determine the threshold value of *StepSize* above which instability results, without the use of the weighting function.

By letting μ vary upward in successive calls, it will be found that the following call, with μ = 1, shows a general borderline stability; values of μ above 1.0 (with no weighting function) will clearly diverge. Figure 1.5 shows this condition, in which successive estimates of x lie at approximately ± 4.

LVxGradientViaCP(4,1,0.01,0,0.0001,50)

Example 1.5.   Demonstrate instability using the script *LVxGradientViaCP* with a value of *StepSize* greater than about 0.0588 and a weighting function.

The call

LVxGradientViaCP(4,0.0589,0.01,1,0.0001,20)

will show that the algorithm rapidly diverges when μ is greater than about 0.058, and the fourth argument calls for *CostFcn* to be applied as a weighting function. The result is shown in Fig. 1.6.

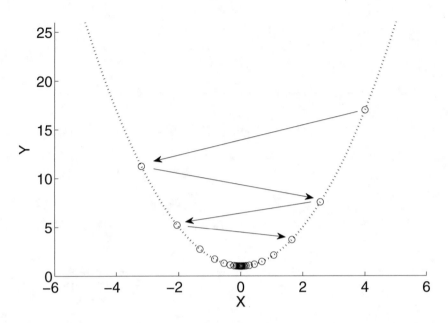

Figure 1.4: A stable overshoot (or underdamped) condition, in which, despite the fact that step size (μ) is over-large, convergence nonetheless occurs since μ is not quite large enough to lead to divergence.

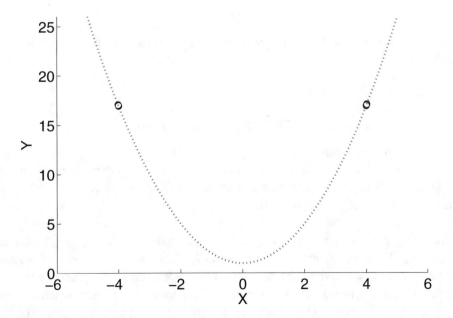

Figure 1.5: A condition in which μ is such that x neither converges nor diverges, but oscillates between two values.

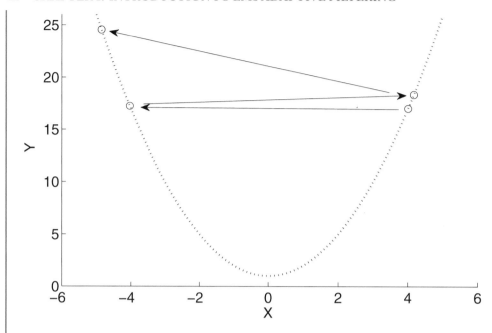

Figure 1.6: A condition in which a value of μ of 0.0589 is sufficient to make the algorithm rapidly diverge. The successive values are connected by arrows; the fifth and successive estimates of x no longer fit on the graph.

Example 1.6.   Demonstrate rapid, stable convergence by use of the weighting function and a moderate value of *StepSize*, 0.03.

A call meeting the stated requirement is

LVxGradientViaCP(4,0.03,0.01,1,0.0001,20)

and the results are shown in Fig. 1.3, plot (d).

The difference from Fig. 1.3, plot (b), in which μ was 0.03, but with no weighting function, is remarkable. In Fig. 1.3, plot (d), the conditions are just right for rapid convergence; in fact, after the second iteration, the movement of x is miniscule since it has very nearly reached the minimum, and the weighting function has a very small value, 1.003, as opposed to its value of about 17 when $x = 4$. Thus the magnitude of the update term was, considering only the value of the weighting function, about 17 times greater at Iteration 1 than it was for Iteration 2, when x had jumped to a value near to zero, the minimum of the cost function. The net value of the update term is the product not only of the weighting function and μ, but also the value of the partial derivative of the cost function with respect to the (single) variable x. It should be intuitively obvious that for an equal change in x (Δx), the cost function changes much more at, say, $x = 4$ than it does when x equals (say) 1. In fact, the partial derivative of our particular cost function is really just the slope of the cost function curve, which of course changes with x. The slope of our *CostFcn* (which equals $x^2 + 1$) is readily obtained from calculus as simply its derivative:

$$\frac{d(CF)}{dx} = \frac{d(x^2 + 1)}{dx} = 2x \tag{1.9}$$

Thus we can compute the value of the coefficient update term at different values of x to see how it varies with x. The basic coefficient update formula is

$$\Delta C = \mu \left(\partial(CF)/\partial(x[n])\right)CF[n] \tag{1.10}$$

with

$$\frac{\partial(CF)}{\partial(x[n])} = \frac{d(CF)}{dx} = 2x$$

For $x = 4$, $\mu = 0.03$, and $CF[4] = 17$ ($4^2 + 1 = 17$), we get

$\Delta C = (0.03)(2)(4)(17) = 4.08$

$x[2] = x[1] - \Delta C$

$x[2] = 4 - 4.08 = -0.08$

This places $x(2)$ near the minimum of the cost function. If you run the script with these conditions, the actual numbers will vary according to the amount of noise you specify in the function call. It should be clear that if the initial value of x is too large, the initial update term will take the next estimate of x to a value having a magnitude greater than the initial value of x, but with opposite sign. The third value of x will have an even larger magnitude, with opposite sign from the second value of x, and so forth.

- Since we have an explicit formula for the cost function, it is possible to compute values for the variables, such as the initial value of x, μ, etc., which will result in borderline stability, or an immediate jump to the minimum value of the cost function. In most real problems, of course, no such explicit formula for the cost function is available, but this does not present a problem since a number of methods, including coefficient perturbation, are available for estimating the gradient and thus enabling the algorithm to navigate to the bottom of the performance surface.

- The optimization and control of μ (*StepSize*) is an important consideration in adaptive filtering. If μ is too small, convergence is very slow, and if it is just right, a single iteration may place the independent variable very near the optimum value, i.e., where the cost function is minimized. A weighting function can be used to effectively boost the value of μ when the distance from the cost function minimum is great, and reduce it as the cost function is reduced. If μ is larger than optimum, overshoot occurs: x overshoots the target. In fact, if μ is too large, the algorithm will diverge.

- Another reason to reduce μ near convergence is that the steady-state error or misadjustment is larger if μ is larger.

- μ is usually controlled by a weighting function, which must at least maintain stability, and hopefully allow for good convergence speed.

- The weighting function chosen for the script *LVxGradientViaCP* is unregulated, meaning that account has not been taken for all input conditions. For example, the call

$$\textbf{LVxGradientViaCP(400,0.03,0.01,1,0.0001,75)}$$

causes divergence because the weighting function is used with a very large starting value of $x = 400$, which causes the initial update term (considering the value of the weighting function, which is 16,001) to be too large. It should be obvious that the size of the update term must be carefully controlled to maintain adequate convergence speed and stability. We will discuss methods for accomplishing this in practical algorithms, such as the LMS algorithm, and a regulated version called the NLMS algorithm, later in the chapter.

1.7   TWO VARIABLE PERFORMANCE SURFACE

So far, we have examined adaptive problems with just one independent variable. A problem which has two variables is that of fitting a straight line to a set of points that may be distributed randomly in a plane, but which may be modeled by the point-slope form of the equation for a line in the x-y plane:

$$y = Mx + b$$

where M is the slope of the line, and b is its y-axis intercept.

The coefficient perturbation method of gradient estimation is employed in the script

$$LVxModelPtswLine(TestM, TestYInt, xTest,...$$

$$Mu, yMu2MuRatio, MStart, yIntStart, NoIts)$$

a typical call for which would be:

$$\textbf{LVxModelPtswLine(2,0,[-10:1:10],0.005,1,-10,8,40)} \tag{1.11}$$

the results of which, plotted on the performance surface, are shown in Fig. 1.7.

This script allows you to specify the slope and y-intercept values (*TestM* and *TestYInt*) which are used to compute the test points to be modeled. *Mu* is the parameter μ, which scales the update term. To enable speedier convergence, the value of μ to be used for the y-intercept variable is equal

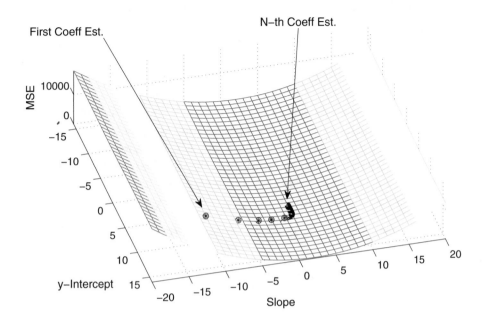

Figure 1.7: Performance Surface with plot of successive coefficient estimates using equal step size weights for both variables in coefficient update, leading to very slow movement down the Performance Surface once the bottom of the gently-sloped trough portion has been reached. This illustrates the Method of Steepest Descent.

to *Mu* multiplied by *yMu2MuRatio*. The fact that different variables have, in general, different optimum step sizes, is an important point that we'll discuss and experiment with below.

MStart and *yIntStart* specify the starting values of the variables. You can experiment with different initial values of the variables to see what happens. In some cases, when an algorithm is being used in specific environments, which may change suddenly but in a historically known way, the values of the variables may be preset to values near what they are likely to eventually converge to, based on prior experience. This technique can greatly reduce convergence time.

Figure 1.8 shows a line to be modeled and the first, second, and fiftieth iterations of the modeling process. The process is very straightforward. The line to be modeled is created using given slope (*M*) and y-Intercept (*B*) values. Since discrete mathematics are used, a set of test values are determined, starting with a range of values of *x*, such as

$$x = [0 : 0.5 : 3]$$

for example. A set of corresponding values of *y* is generated using the vector *x* and the given values of *M* and *B*. The first estimates for *M* and *B*, which we'll call *eM* and *eB*, are used to create, using the x vector, a corresponding set of y values, which we can call eY. Then the mean squared error is given as

$$MSE = \frac{1}{N} \sum_{n=0}^{N-1} (y[n] - eY[n])^2$$

From this, estimates of the partial derivative of MSE with respect to eM and eB (the current estimated values of M and B) are made using coefficient perturbation using central differences, and then new values for eM and eB are computed using the relationship specified by Eq. (1.8). The process is repeated until some criterion is reached, usually that the magnitude of the error becomes quite small, or that the amount of improvement per iteration becomes less than some threshold value, or perhaps some combination of the two.

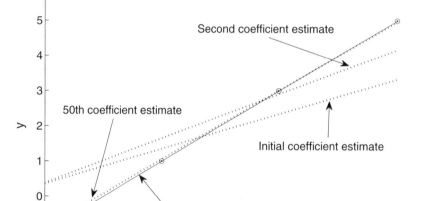

Figure 1.8: Determining the slope and y-Intercept of a line using an iterative method. The line to be modeled is shown as a solid line which passes through the sample or data points which are plotted as circles. Also shown are (estimated) dashed lines based on the first, second, and 50th coefficient estimates.

Figure 1.8 was created using the script

$$LV\_ModelLineLMS(xVec, yVec, Mu, Its)$$

and in particular, the call

LV_ModelLineLMS([0,1,2,3], [-1,1,3,5], 0.05, 50)

On the other hand, the script *LVxModelPtswLine* provides a separate value of μ for the y-intercept variable by weighting μ for the y-intercept variable relative to μ for the slope variable according to the user-input variable *yMu2MuRatio*. It is easy to see why the coefficient for slope converges more quickly than the coefficient for y-intercept by inspecting Fig. 1.8 and imagining how much a small amount of misadjustment of slope affects the value of MSE compared to an equal amount of misadjustment of y-intercept. Clearly, a slight misadjustment of slope leads to a much larger change in MSE than does the same misadjustment in y-intercept. By artificially boosting the coefficient update term weight for y-intercept, we can make the algorithm converge more quickly.

Let's return to Fig. 1.7, which shows the coefficient estimate track resulting from making the call to the script given at (1.11), plotted with respect to the performance surface, which was itself generated by systematically varying the two coefficients, computing the resulting MSE, and then plotting the resultant data as a surface. Notice in Fig. 1.7 that the performance surface has different slopes in the two dimensions representing slope and y-intercept. Since there are two independent variables, each must converge to its ideal value to minimize the cost function. Since the partial derivative of one of them is very much smaller than the other (at least starting from the initial coefficient estimate, neither coefficient of which is equal to its ideal or test value), it will actually converge much more slowly. In this case, *yIntNow* (the variable representing the estimate for y-intercept in the script) converges very slowly compared to *SlopeNow* (the variable representing the estimate for slope in the script).

The method of steepest descent, which uses the negative of the gradient to estimate the next point to move to on the performance surface, moves along the direction which reduces the MSE by the largest amount for a given very small amount of movement. In Fig. 1.7, the performance surface is much like a river that has relatively steep banks which slope not only down to the river, but which slope slightly downward along the direction of river flow. If a ball were released from the top of such a river bank, its direction of travel would be determined by gravity. It would appear to mostly move directly toward the river, i.e., straight down the river bank to the water. There would, however, under the given assumptions, be a slight bias toward the direction the river is flowing since in fact the banks also have a slight downward slope in the same direction as the river's flow. In fact, the ball would be following gravity along a direction which reduces gravitational potential the greatest amount for a given amount of travel. The path a ball would follow down such a river bank is similar to the path followed in Fig. 1.7 by the coefficient estimate pair *yIntNow* and *SlopeNow* as they move toward the correct values of *TestYInt* and *TestM*. Carrying the analogy further, if the point of lowest gravitational potential (analogous to the minimum value of the cost function) is at the river's mouth, the fastest way to get there (along the shortest path) is one which heads directly for the mouth, not one which first rolls down the banks to the river and then suddenly turns downstream to head for the mouth at a very leisurely pace (this particular situation is analogous to the one shown in Fig. 1.7).

- **The method of steepest descent is not necessarily one which takes the shortest and/or fastest path to the minimum of the cost function. This is a function of the shape of the performance surface.**

Method of Steepest Descent Modified

In the script *LVxModelPtswLine*, the rate at which the slower coefficient converges is modified by the multiplier *yMu2MuRatio*, leading to a more optimal path from starting point to minimum value of cost function.

Example 1.7.   Experimentally determine a value of *yMu2MuRatio* which will facilitate rapid convergence of the algorithm in the script LVxModelPtswLine.

A good way to start is with a basic call that results in stable convergence, and to gradually increase the value of *yMu2MuRatio*. A little such experimentation results in a useful set of parameters for convergence with *Mu* = 0.009 and *yMu2MuRatio* = 37. The result of the call

$$\text{LVxModelPtswLine(2,0,[-10:1:10],0.014,37,-10,8,12)}$$

is shown in Fig. 1.9. You should repeat the above call a number of times, varying the parameters *TestM*, *TestYInt*, *MStart*, and *yIntStart*, to observe the result. Does the algorithm still converge immediately?

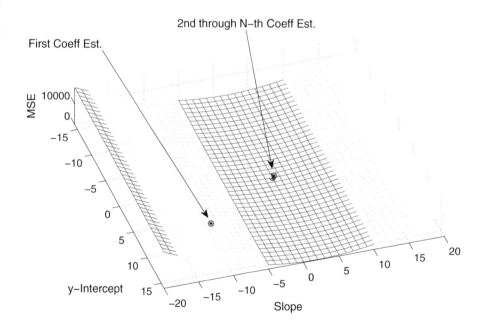

Figure 1.9: Performance Surface with plot of successive coefficient estimates utilizing a weighting function to accelerate the slower-converging of the two coefficients. Note that convergence is extremely rapid since the value of μ for the y-intercept variable has been adjusted to greatly speed the convergence of that variable toward its value of best adjustment.

Using the river analogy, we can see that instead of moving the way gravity would dictate, which would be along a nearly direct path toward the river (with only a slight bias downstream), and then suddenly turning to head directly downstream, the algorithm immediately headed directly for the point of minimized cost function, (the river mouth in our analogy).

In Fig. 1.9, near-convergence occurs in a few iterations, with additional iterations making only very minute improvements.

1.8   AN IMPROVED GRADIENT SEARCH METHOD

Thus far in the chapter we have used coefficient perturbation to estimate the gradient in several adaptive processes; we will now develop an analytic expression to estimate the gradient in the two-variable line-modeling process just discussed. This will be accomplished by writing an analytic expression for the mean squared error, and then obtaining the partial derivative with respect to m and b. This will permit us to write coefficient update equations consisting of analytic expressions.

If we represent the points to be modeled as two vectors $xVec$ and $yVec$ of length M, we can write the mean squared error as

$$MSE[n] = \frac{1}{M} \sum_{i=0}^{M-1} (yVec[i] - (m[n])(xVec[i]) - b[n])^2 \tag{1.12}$$

which is the sum of the mean squared error at each point be modeled. Since $yVec$ and $xVec$ are constants, and b is considered a constant when computing the partial derivative of MSE with respect to m, we get the partial derivative of MSE with respect to m as

$$\frac{\partial(MSE[n])}{\partial(m)} = \frac{2}{M} \sum_{i=0}^{M-1} (yVec[i] - (m[n])(xVec[i]) - b[n])(-xVec[i]) \tag{1.13}$$

and similarly the partial derivative of MSE with respect to b as

$$\frac{\partial(MSE[n])}{\partial(b)} = \frac{2}{M} \sum_{i=0}^{M-1} (yVec[i] - (m[n])(xVec[i]) - b[n])(-1) \tag{1.14}$$

Note that the scalar error $E[i, n]$ at data point (x_i, y_i) at sample index n is

$$E[i, n] = yVec[i] - (m[n])(xVec[i]) - b[n]$$

so we can rewrite Eqs. (1.13) and (1.14) as

$$\frac{\partial(MSE[n])}{\partial(m)} = -\frac{2}{M} \sum_{i=0}^{M-1} E[i, n](xVec[i]) \tag{1.15}$$

$$\frac{\partial(MSE[n])}{\partial(b)} = -\frac{2}{M} \sum_{i=0}^{M-1} E[i, n] \tag{1.16}$$

The corresponding update equations are

$$m[n + 1] = m[n] - \mu \frac{\partial(MSE[n])}{\partial(m)} \tag{1.17}$$

and

$$b[n + 1] = b[n] - \mu \frac{\partial(MSE[n])}{\partial(b)} \tag{1.18}$$

The above formulas are embodied in m-code in the script (see exercises below)

$$LVxModelLineLMS\_MBX(M, B, xVec, Mu, bMu2mMuRat, NoIts)$$

where M and B are values of slope and y-intercept, respectively, to be used to construct the y values to be modeled from $xVec$; Mu (μ) and $NoIts$ have the usual meaning. The value of Mu is controlled for stability (see exercises below). The rate of convergence of b is much slower than that of m, and as a result the script receives the input variable $bMu2mMuRat$, which is used to generate a boosted-value of Mu for use in the b-variable update equation (1.18). The following call leads to slow, stable convergence:

$$\textbf{LVxModelLineLMS\_MBX(3,-1,[-10:1:10],0.5,1,150)} \tag{1.19}$$

On the other hand, after determining a suitable value by which to boost μ for the b-coefficient update term, the same thing can be accomplished in ten iterations rather than 150:

$$\textbf{LVxModelLineLMS\_MBX(3,-1,[-10:1:10],0.5,29,10)}$$

1.9   LMS USED IN AN FIR

1.9.1   TYPICAL ARRANGEMENTS

In order to visualize various aspects of the LMS algorithm as applied to an FIR, we'll use a 2-tap FIR, which allows the graphing of both coefficient values versus MSE.

Figure 1.10 shows a basic 2-Tap adaptive FIR filter which is to model an LTI system represented by the box labeled "Plant." For continuous domain systems, the interfacing to and from the LMS adaptive filter system, is, of course, managed with ADCs and DACs, operated at a suitable sample rate with the required anti-aliasing filter. In most books, such ADCs and DACs are generally not shown, but are understood to be a necessary part of an actual working embodiment.

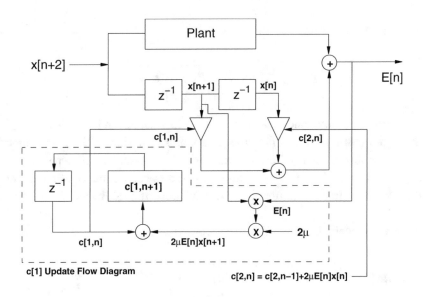

Figure 1.10: A flow diagram of a 2-Tap FIR configured to model a Physical Plant, which presumably has an impulse response of one or two samples length, i.e., a total delay equal or less than that of the adaptive FIR, which must be able to match each delay in the Plant's impulse response to model it. The Plant's impulse response is thus being modeled as a 2-Tap FIR having coefficients PC_1 and PC_2.

In the example shown in Fig. 1.10, the Plant is itself a 2-tap FIR, with unknown tap weights. The LMS adaptive filter's task will be to model the Plant, i.e., determine its impulse response, which consists of the two tap weights.

In general, modeling a particular Plant requires that the FIR have a length at least equal to the longest time delay experienced by the signal travelling through the Physical Plant. In a real-world system, of course, the number of taps involved might be dozens, hundreds, or even thousands, depending on the particular application. A 2-Tap example, however, allows graphing of the coefficients versus MSE or Scalar Error Squared (a cost function which we will investigate shortly), which allows the reader to attain a visual impression of the manner in which the adaptive filter works.

Referring to Fig. 1.10, the test signal enters at the left and splits, one part heading into the Plant, and the other part entering the filter's two series-connected delay stages (labeled z^{-1}). The signal at the output of each delay stage goes to a coefficient multiplier.

1.9.2   DERIVATION

The system of Fig. 1.10 may be represented as follows:

$$
\begin{aligned}
Err[n] \quad = \quad & (PC_1 \cdot x[n+1] + PC_2 \cdot x[n])... \\
& -(c_1 \cdot x[n+1] + c_2 \cdot x[n])
\end{aligned}
$$

where PC means *Plant Coefficient* and c_1 and c_2 are the filter tap coefficients. In the earlier examples, we used Mean Squared Error (MSE) as a cost function. This works well when MSE can be directly obtained or computed. For many problems, however, such is not the case. For the type of system shown in Fig. 1.10, the Plant's impulse response is unknown, and the consequence of having only the Plant's output to work with is that we do not have a true measure of MSE by which to estimate the gradient.

A true MSE measure of coefficient misadjustment would be

$$
\begin{aligned}
TrueMSE[n] \quad = \quad & ((PC_1 - c_1[n]) \cdot x[n+1])^2 + ... \\
& ((PC_2 - c_2[n]) \cdot x[n])^2
\end{aligned}
$$

This cost function has a true global or unimodal minimum since it can only be made equal to zero when both coefficients are perfectly converged to the plant values.

In accordance with our previously gained knowledge, we can write a general formula for the partial derivative of *TrueMSE* with respect to the ith tap coefficient as

$$
\frac{\partial(TrueMSE[n])}{\partial(c_i[n])} = -2(PC_i - c_i[n]) \cdot (x[n+N-i]) \tag{1.20}
$$

where PC_i means the ith Plant Coefficient, c_i is the ith tap coefficient, N is the total number of taps, and n is the current sample being computed (and remember that the partial derivative of MSE for any Plant coefficient is zero), and the partial derivative of MSE for any given coefficient $c_j[n]$ with respect to coefficient $c_i[n]$ is always zero.

The gradient might be notated as

$$
\nabla(TrueMSE[n]) = -2 \sum_{i=1}^{N} (PC_i - c_i[n]) \cdot x[n+N-i]) \cdot \hat{c}_i \tag{1.21}
$$

Equation (1.21) requires explicit knowledge of the scalar error contributed by each pair of plant and filter coefficients. Returning to Eq. (1.20), what can readily be seen is that the partial derivative for each coefficient is dependent on its individual misadjustment relative to the equivalent plant coefficient, as well as the value of the signal x at the tap, which changes with n. The performance surface for a true-MSE cost function is quadratic; for a two-tap filter, it graphs as a paraboloid of revolution, as shown in Fig. 1.11, in which MSE in decibels is graphed vertically against the two coefficients.

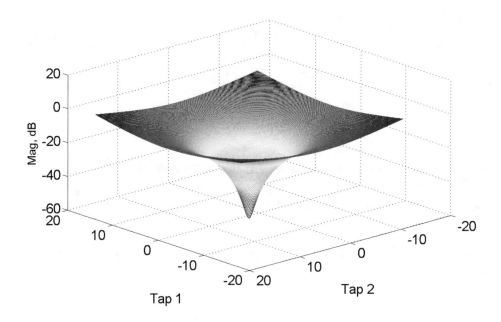

Figure 1.11: Performance surface for a 2-Tap Adaptive filter using a True-MSE algorithm.

The scalar error, the difference between the plant output and the filter output, is all we have access to in most problems, and, for the two-tap problem at hand, we must approximate the true MSE by squaring the scalar error:

$$ApproxMSE[n] = ScalarError^2 =$$

$$((PC_1 - c_1[n])x[n+1] + (PC_2 - c_2[n])x[n])^2 \tag{1.22}$$

We can derive an estimate for the partial derivative of *ApproxMSE* with respect to, say, c_1, (as in previous examples) as

$$\frac{\partial(ApproxMSE[n])}{\partial(c_1[n])} = 2((PC_1 - c_1[n])x[n+1]...$$

$$+ (PC_2 - c_2[n])x[n])(-x[n+1]) \tag{1.23}$$

This formula may be generalized for an N-Tap FIR as

$$\frac{\partial(Err[n]^2)}{\partial(c_k[n])} = 2\left[\sum_{i=1}^{N}(PC_i - c_i[n])x[n+N-i]\right](-x[n+N-k]) \tag{1.24}$$

which may be rewritten as

$$\frac{\partial (Err[n]^2)}{\partial (c_k[n])} = -2E[n](x[n + N - k]) \tag{1.25}$$

where $E[n]$, the scalar error at sample time n, is

$$E[n] = \sum_{i=1}^{N} (PC_i - c_i[n])x[n + N - i]$$

Equation (1.25) may be interpreted as saying that the partial derivative of the scalar error squared at sample time n with respect to the kth tap coefficient at sample time n is negative two times the scalar error at sample time n, $E[n]$, multiplied by the signal value at the kth tap at sample time n.

- The coefficient update algorithm for an LMS adaptive FIR is, after inserting the update term weighting factor μ

$$c_i[n + 1] = c_i[n] + 2\mu \cdot E[n] \cdot x_i[n] \tag{1.26}$$

where $c_i[n]$ represents the tap coefficient of tap index i and iteration n, μ is a scalar constant which scales the overall magnitude of the update term, $E[n]$ represents the scalar error at iteration n, and $x_i[n]$ is the signal value at tap i at iteration n.

- While the LMS algorithm is elegant, and moderately robust, it is not as robust as a true-MSE algorithm in which the gradient is estimated from knowledge of the actual MSE. Equating squared scalar error to MSE allows the simplicity and usefulness of the LMS algorithm, but there are limitations on the frequency content of the input signal to ensure that the LMS algorithm properly converges. This will be explored below.

1.9.3   LIMITATION ON Mu

One idea of the maximum value that the stepsize Δ (2μ in Eq. (1.26), for example) may assume and remain stable is

$$0 < \Delta < \frac{1}{10NP} \tag{1.27}$$

where N is the filter length and P is the average signal power over M samples of the signal:

$$P = \frac{1}{M} \sum_{i=0}^{M-1} (x[i])^2$$

Thus the quantity NP is a measure of signal power in the filter. The script (see exercises below)

$$LVxLMSAdaptFiltMuCheck(k, NoTaps, C, SigType, Freq, Delay)$$

filters either random noise of standard deviation k, or a cosine of amplitude k and frequency zero (DC), Halfband, Nyquist, or $Freq$, with an LMS adaptive FIR having a length equal to $NoTaps$, and plots the test signal in one subplot, and the filter output in a second subplot. The stepsize for coefficient update is computed as the high limit given by Eq. (1.27), but is also multiplied by the constant C for testing purposes. Variation of k and N has little effect on the convergence rate and stability of the algorithm when stepsize is computed using Eq. (1.27). In practice, a value of stepsize considerably larger may often be used to achieve convergence in a number of iterations approximately equal to ten times the filter length. The parameter C can be experimentally used to observe this. Figure 1.12 shows the error signal for various values of C from 1 to 10 with $k = 1$ and $NoTaps = 100$, white noise as the test signal, and $Delay = 10$ (see the exercises below for a complete description of the input arguments). Figure 1.13 shows the same thing as Fig. 1.12 except that $k = 3$ and $NoTaps = 40$. For the examples shown, the best value of C appears to lie between 5.0 and 7.0-this value range allows for both speedy convergence and good stability.

1.9.4   NLMS ALGORITHM

A standard method to assist in controlling the stepsize in Eq. (1.26) is to divide by a factor related to the signal power in the filter, to which is added a small number ϵ in case the signal power in the filter should be zero. We thus get

$$c_i[n+1] = c_i[n] + \frac{\Delta}{\sum x_i[n]^2 + \epsilon} \cdot Err[n] \cdot x_i[n]$$

(where $\Delta = 2\mu$) and refer to this as the Normalized Least Mean Square, or NLMS, algorithm. Note that in Eq. (1.27), the power is averaged over the entire input signal. In the NLMS method, only signal values in the filter are taken into account. The NLMS is thus more adaptive to current signal conditions. The various scripts we will develop using the LMS algorithm in an FIR will generally be NLMS algorithms.

The script (see exercises below)

$$LVxLMSvNLMS(kVec, NoTaps, Delta, Mu, SigType, Freq, Delay)$$

passes a test signal determined by the value of $SigType$ through two FIRs in parallel, one FIR using the LMS algorithm with a stepsize fixed as Mu, and the other FIR using the NLMS algorithm having a stepsize equal to $Delta$ divided by the signal power in the filter (plus a very small constant to avoid potential division by zero). The test signal is given an amplitude profile defined by $kVec$, which specifies a variable number of different equally spaced amplitudes imposed on the test signal.

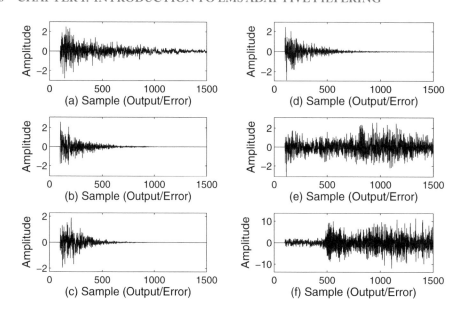

Figure 1.12: (a) Error signal for Stepsize = $C/(10NP)$, with $C = 1$, $k = 1$ (see text); (b) Same with $C = 3$; (c) Same with $C = 5$; (d) Same with $C = 7$; (e) Same with $C = 9.5$; (f) Same with $C = 10$.

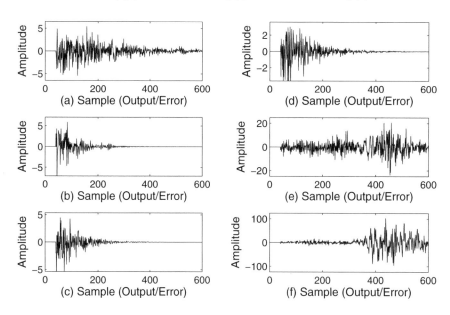

Figure 1.13: (a) Error signal for Stepsize = $C/(10NP)$, with $C = 1$, $k = 3$ (see text); (b) Same with $C = 3$; (c) Same with $C = 5$; (d) Same with $C = 7$; (e) Same with $C = 9.5$; (f) Same with $C = 10$.

In this manner, the test signal, be it noise or a cosine, for example, can undergo step changes in amplitude that test an algorithm's ability to remain stable.

The call, in which Mu has been manually determined to allow the LMS algorithm to converge,

$$\text{LVxLMSvNLMS}([1,5,15,1],100,0.5,0.00005,1,[],50)$$

results in Fig. 1.14, while the call

$$\text{LVxLMSvNLMS}([1,5,25,1],100,0.5,0.00005,1,[],50)$$

in which the test signal amplitude has been boosted in its middle section, results in Fig. 1.15, in which it can be seen that the LMS algorithm has become unstable and divergent, while the NLMS algorithm has remained stable and convergent.

Limitations of the LMS/NLMS Algorithms

We now proceed to study the LMS algorithm in detail for the 2-tap FIR example and observe its behavior with different types of input signals. Referring to Eq. (1.22), the cost function represented by this equation does not have a single minimum; rather, the performance surface (which resembles the lower half of a pipe split along its longitudinal axis) is linear in one direction, and quadratic in a direction perpendicular to the first direction; it has a perfectly flat bottom and defines (along that bottom) an infinite number of pairs of coefficients (c_1, c_2) which will make *ScalarErrorSquared* = 0.

We can derive an expression for this type of performance surface by determining what values of c_1 and c_2 will drive the cost function (*ApproxMSE*) to any arbitrary value ESq. Then, from Eq. (1.22) we get

$$(PC_1 - c_1[n]) \cdot x[n+1] + (PC_2 - c_2[n]) \cdot x[n] = \sqrt{ESq}$$

We proceed by solving for $c_2[n]$ in terms of $c_1[n]$:

$$c_2[n] = -(\frac{x[n+1]}{x[n]}) \cdot c_1[n] + ...$$

$$(PC_2 + PC_1 \cdot \frac{x[n+1]}{x[n]} - \frac{\sqrt{ESq}}{x[n]}) \tag{1.28}$$

Equation (1.28) tells, for any $c_1[n]$, what value of $c_2[n]$ is necessary to force the scalar error to be $\sqrt{ESq}$ (or scalar error squared to be ESq). Equation (1.28) is a linear equation of the familiar form $y = Mx + b$ (or, in the iterative or adaptive sense, $y[n] = M[n]x[n] + b[n]$ in which $y[n]$ equates to $c_2[n]$,

$$M[n] = -\frac{x[n+1]}{x[n]} \tag{1.29}$$

and

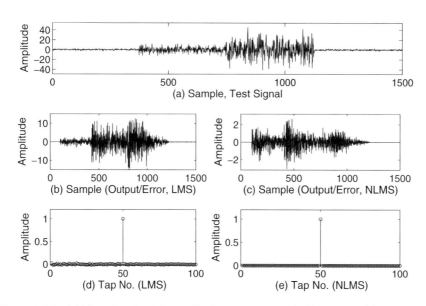

Figure 1.14: (a) Test signal, with amplitude segments scaled by $kVec$; (b) Error signal from the LMS algorithm; (c) Error signal from the NLMS algorithm; (d) Final Tap Weights for the LMS algorithm; (e) Final Tap Weights for the NLMS algorithm.

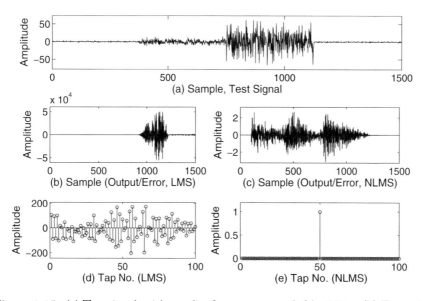

Figure 1.15: (a) Test signal, with amplitude segments scaled by $kVec$; (b) Error signal from the LMS algorithm; (c) Error signal from the NLMS algorithm; (d) Final Tap Weights for the LMS algorithm; (e) Final Tap Weights for the NLMS algorithm.

$$b[n] = (PC_1 + PC_0 \cdot \frac{x[n+1]}{x[n]} - \frac{\sqrt{ESq}}{x[n]}) \tag{1.30}$$

It can be seen that the slope $M[n]$ in Eq. (1.29) is just the ratio of the two samples in the filter. If the input signal is changing, then the slope will also change at each sample time, as will the value of $b[n]$ in Eq. (1.30).

However, no matter what the slope M and intercept value b are, the line defined by the equation always passes through the point of perfect coefficient adjustment, that is, $c_1 = PC_1$ and $c_2 = PC_2$.when $ESq = 0$. This can be readily verified by setting $c_1 = PC_1$ and $ESq = 0$ in Eq. (1.28).

The performance surface generated by Eq. (1.28) is linear in a first direction (lines of equal ESq are parallel to this first direction), and quadratic in the direction perpendicular to the first direction. We'll refer to this performance surface as a *Linear-Quadratic Half Pipe* (the surface is more pipe-like in appearance when magnitude is graphed linearly), and to lines of equal squared error as *IsoSquErrs*. Figure 1.16 shows an example of an LQHP performance surface, Scalar Error squared in dB is plotted against the coefficients.

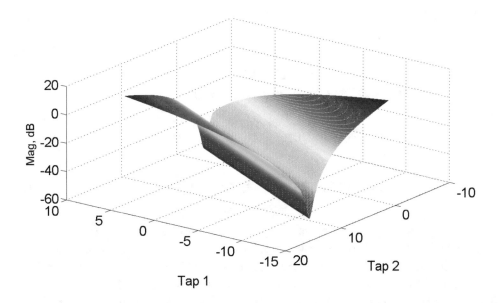

Figure 1.16: Performance surface for a 2-Tap LMS Adaptive filter for a particular instantaneous pair of signal values in the filter; orientation of the "trough" in the performance surface changes as the ratio of the two signal values changes.

There are several important things to note about this:

- The (negative of the) gradient will not, in general, be headed toward the point of best coefficient adjustment, but rather it will be headed along the shortest path (a normal line, possibly corrupted with noise) toward the bottom of a LQHP which varies orientation in the (c_1, c_2) plane (slope- or angle-wise) according to the current values of the input signal in the filter.

- The very bottom of each of the these LQHPs actually passes through the point of ideal coefficient adjustment when $ESQ = 0$, no matter what the slope or angle is of the LQHP in the (c_1, c_2) plane.

- Accordingly, the task of getting to the ideal adjustment point (i.e., when the coefficient estimates properly model the Plant) depends on the orientation of LQHPs changing from one iteration to the next.

- This process may be viewed graphically by calling the script

$$LV\_FIRLMSPSurf(PC1, PC2, Mu, c1St, c2St, ...$$
$$NoIts, tstDType, SineFrq, NoiseAmp)$$

a typical call for which would be

$$\textbf{LV\_FIRLMSPSurf(1,-0.5,0.35,5,6,12,0,[],0)} \tag{1.31}$$

The input argument *tstDType* is an integer that selects the type of test sequence, which allows testing of the algorithm with several possible data types. Use *tstDType* = 0, for white noise; *tstDType* = 1, for a unit step, or DC signal; use *tstDType* = 2 for a sequence of alternating 1's and -1's, representing a sine wave at the Nyquist limit; use *tstDType* = 3 for a test signal at the half-band frequency ([1,0,-1,0,1,0,-1,0 ...]), and finally, a value of *4* yields a pure (noiseless) sine wave, the frequency of which may optionally be specified by entering another number after *tstDType*–if in this case *SineFrq* is passed as the empty matrix, the sine's frequency defaults to 25 cycles/1024 samples. If you aren't calling for a sine wave as the test signal, you may pass *SineFreq* as the empty matrix, [ ], as shown in the sample call above.

A coefficient track which resulted from the call at (1.31), which specifies the test signal as random noise, is shown in Fig. 1.17.

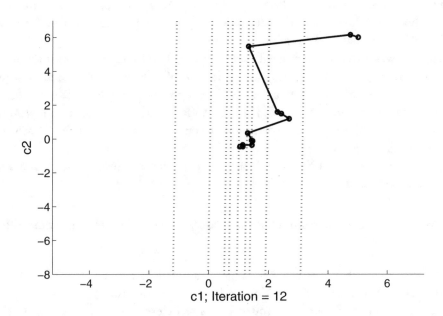

Figure 1.17: Coefficient track for a 2-Tap LMS Adaptive filter using random noise as the test signal. Minimum of cost function is a line (the center-most dotted line in the plot) rather than a point. A random noise test signal causes the orientation of the performance surface to change randomly, resulting in convergence to the correct coefficient values, which are 1, -0.5.

The script

$$LV\_FIRLMS6Panel(PC1, PC2, Mu, c1St, c2St, ...$$
$$NoIts, tstDType, SineFrq, NoiseAmp)$$

is similar to the script *LV_FIRLMSPSurf*, except that it plots the most recent six iterations in one figure.

Example 1.8. Compute and plot the performance surface (several IsoSqErrs will do) and coefficient track for a number of iterations of an adaptive LMS 2-tap FIR, with white noise as the signal.

Figure 1.18, which was produced by the script call

$$\textbf{LV\_FIRLMS6Panel(-2,3,0.25,0,0,6,0,[ ],0)} \tag{1.32}$$

allows one to see the sequence of events in which the performance surface constantly shifts with each sample, but the coefficient values are inexorably drawn toward proper adjustment–provided certain conditions are met. We'll see what those conditions are shortly.

In Fig. 1.18, noise was used as the test signal. At Iteration 1, the coefficient pair (c_1, c_2) moved from its initial location at (0,0) to a new location, and the direction of movement was generally toward $IsoSquErr(0)$, i.e., the center-most solid line in each of the six graphs, where $ScalarError^2 = 0$. At each successive iteration, the orientation of $IsoSquErr(0)$ shifts, drawing the coefficients toward it. This process is very efficient at drawing the coefficients toward the point of ideal adjustment or convergence when the input signal is noisy.

Unfortunately, any input signal that does not change enough can lead to stagnation and convergence to an incorrect solution.

Example 1.9. Compute and plot the performance surface and successive coefficient estimates for a 2-tap FIR using a DC signal.

Figure 1.19 shows what happens if a DC signal (unit step) is applied to a 2-Tap LMS adaptive FIR filter. You should note that the coefficients do converge to a certain point (i.e., a certain coefficient pair (c_1, c_2)), and the Scalar Error squared is in fact close to zero. Unfortunately, the converged coefficient values are incorrect. A similar thing may happen with other types of highly regular signals, such as a pure (noise-free) sine wave.

Example 1.10. Compute and plot the performance surface and successive coefficient estimates for a 2-tap FIR filtering a sinusoidal signal of frequency 25.

The script call

LV_FIRLMS6Panel(-1,3,0.25,5,3,6,4,25,0)

in which the seventh and eighth input arguments (4, 25) specify a test signal consisting of a pure sine wave of frequency 25 cycles (over 1024 samples), results in the plots shown in Fig. 1.20. The result shown in Fig. 1.20 again shows convergence to an incorrect solution, with, nonetheless, the square of Scalar Error minimized. Logically, a sine wave does change over time, so it is necessary to perform enough iterations so that several cycles of the sine wave are traversed. In general, though, convergence is protracted and uncertain.

Example 1.11. Compute and plot the error signal and coefficient estimates for a 2-tap FIR that models a Plant defined by Plant Coefficients PC1 and PC2, as shown in Fig. 1.10, using a test signal consisting of a sine wave containing, respectively, small, medium, and large amounts of white noise.

The computations may be performed using the script (see exercises below)

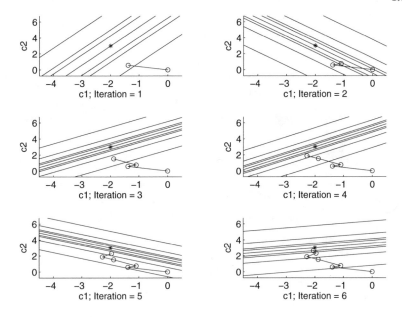

Figure 1.18: Coefficient track in six successive panels/iterations for the two coefficients of an LMS Adaptive filter using random noise as a test signal.

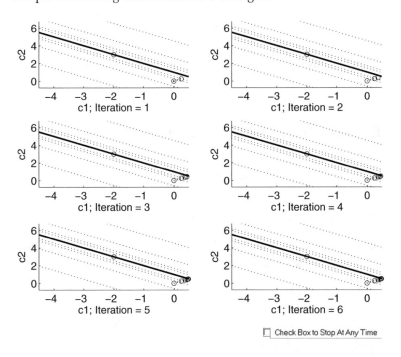

Figure 1.19: Coefficient track in six successive panels/iterations for the two coefficients of an LMS Adaptive filter using a Unit Step (DC) as a test signal.

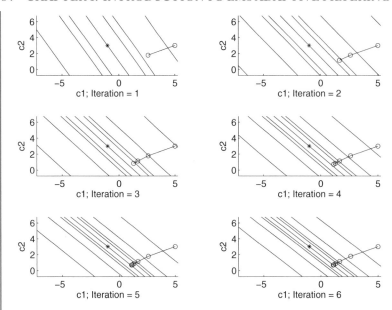

Figure 1.20: Coefficient track in six successive panels/iterations for the two coefficients of an LMS Adaptive filter using a 25 Hz sine wave as a test signal.

$$LVxLMS2TapFIR(PC1, PC2, NoIts, tstDType, CosFrq, TwoMu, NoiseAmp)$$

with the specific calls

LVxLMS2TapFIR(-3.5,1.5,32,5,0.5,0.5,0)

LVxLMS2TapFIR(-3.5,1.5,32,5,0.5,0.5,0.2)

LVxLMS2TapFIR(-3.5,1.5,32,5,0.5,0.5,4)

which yield, respectively, Fig. 1.21, Fig. 1.22, and Fig. 1.23. Note that only the third call, (which uses as a test signal a sine wave to which has been added a large amount of noise) results in rapid convergence to the correct coefficients.

1.10   CONTRAST-TRUE MSE

We have seen that the LMS algorithm, as applied to a typical system, uses the square of scalar error as an estimate of the true MSE in order to estimate the gradient. As a result, the actual

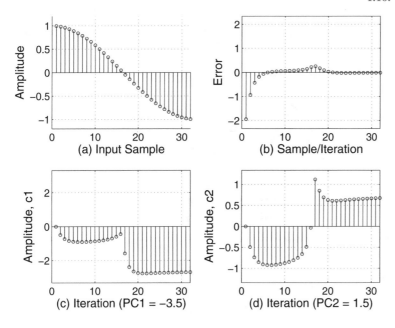

Figure 1.21: (a) Input signal, a sine wave containing a small amount of random noise; (b) Output or error signal; (c) Estimate of 1st coefficient versus iteration; (d) Estimate of 2nd coefficient versus iteration.

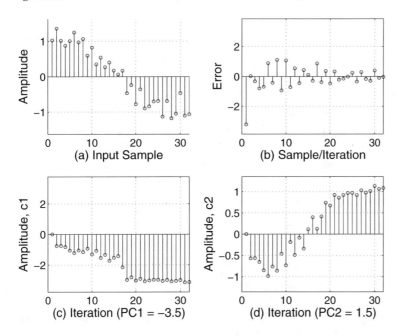

Figure 1.22: (a) Input signal, a sine wave containing a medium amount of random noise; (b) Output or error signal; (c) Estimate of 1st coefficient versus iteration; (d) Estimate of 2nd coefficient versus iteration.

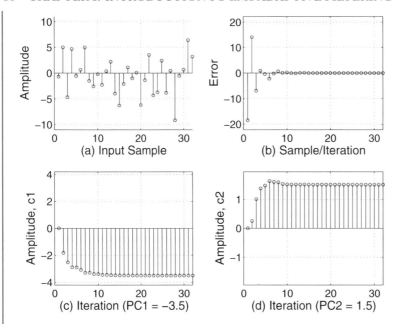

Figure 1.23: (a) Input signal, a sine wave containing a large amount of random noise; (b) Output or error signal; (c) Estimate of 1st coefficient versus iteration; (d) Estimate of 2nd coefficient versus iteration.

performance surface is not unimodal (i.e., does not possess a single minimum at the point of optimum coefficient adjustment). However, under proper conditions, the LMS algorithm behaves as though a true unimodal performance surface were being traversed; it generally performs poorly with input signals having narrow bandwidth spectra, such as DC, low frequency sine waves, etc.

When we actually know the plant coefficients to which the filter is trying to converge, we can construct an LMS-like algorithm that uses this information to give a much better gradient estimate. The result is that the algorithm works well with all signals, including sinusoids, DC, Nyquist, etc. The purpose of using an adaptive filter, of course, is usually to determine the plant coefficients, so the situation hypothesized here is mainly academic. However, it serves to illustrate the difference between a true-MSE cost function and one based on the square of scalar error.

The script

$$LV\_FIRPSurfTrueMSE(PC1, PC2, Mu, c1Strt, ...$$
$$c2Strt, NoIts, tstDType, CosFrq)$$

embodies such an algorithm, where $PC1$ and $PC2$ are the Plant coefficients to be modeled, $c1Strt$ and $c2Strt$ are the initial guesses or starting values for the coefficient estimates, $NoIts$ is the number of iterations to perform, Mu has the usual meaning, and $tstDType$ selects the type of test signal. Pass *tstDType* as 0 for white noise, 1 for a unit step (DC), 2 for the Nyquist Limit frequency, 3 for the half-band frequency, and 4 for a cosine of frequency *CosFrq* (may be passed as [] if a cosine is not being used as a test signal).

Example 1.12. Compute and plot the performance surface for a 2-tap FIR using a true-MSE-based coefficient update algorithm; use a sinusoidal input.

Figure 1.24 shows the results after making the call

LV_FIRPSurfTrueMSE(-1,3,1,5,6,7,4,125)

Even though the test signal employed to generate the coefficient track in Fig. 1.24 was a sine wave, convergence is proper.

Example 1.13. Compute and plot the performance surface for a 2-tap FIR using a true-MSE-based coefficient update algorithm; use a DC input.

The script call

LV_FIRPSurfTrueMSE(-1,3,1,5,6,5,1,[])

uses a DC (unit step) signal as the test signal, and results in a coefficient track which forms a straight line from the entry point to the point of ideal adjustment, as shown in Fig. 1.25.

1.11 LMS ADAPTIVE FIR SUMMARY

In general, a pure frequency (a sinusoid) poses difficulties for an LMS algorithm, which uses the square of the scalar error as the cost function rather than the true MSE. However, depending on exact conditions, it is possible to achieve proper convergence. This is more likely with a higher frequency than with a lower frequency–the key word is change. The more rapid the change in test signal values, especially change in sign, the better. Mixtures of sinusoids, to the extent more rapid sign and amplitude change result, improve performance. Recall from some of the examples, however, that an LMS adaptive filter can converge with a narrow bandwidth input (and in fact minimize the scalar error), just as with other, wider-bandwidth signals. The difference is that with narrow bandwidth signals, the filter coefficients do not necessarily model the Physical Plant properly, since there is no unique set of coefficients that will minimize the error (rather, there are many possible sets of coefficients that can minimize the scalar error).

Hence it is, in general, good to ensure that LMS adaptive filters receive signals for processing which are at least somewhat noisy in nature. In many applications, such as noise cancellation, the input signal actually is more or less pure noise. In some applications, such as, for example, feedback cancellation in a hearing aid, it may be necessary to inject noise into the signal to cause the LMS adaptive algorithm to converge. We'll see an application in the next chapter, using a special filter

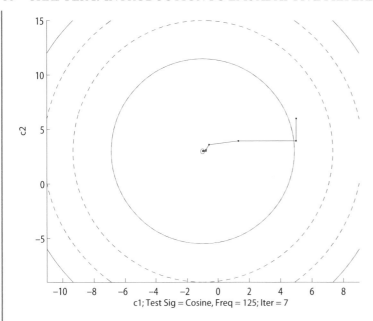

Figure 1.24: Plot of $c1$ and $c2$ for seven iterations of a True-MSE-based gradient algorithm, using as a test signal a cosine of 125 cycles over 1024 samples, i.e., a normalized frequency of a $125/512 = 0.24\pi$. Plant coefficients (-1,3) lie at the center of the plot.

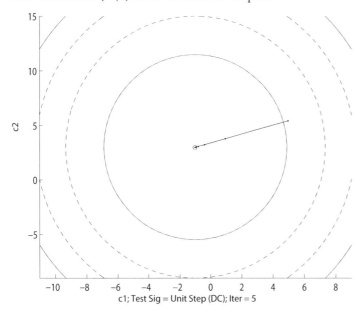

Figure 1.25: Plot of c1 and c2 for 5 iterations of a True-MSE-based gradient algorithm, using as a test signal a 1024 sample unit step. Plant coefficients (-1,3) lie at the center of the plot.

topology, however, in which sinusoidal signals, but not random signals, can be cancelled using an LMS adaptive filter.

We have looked extensively at the inner workings and hidden mechanisms of the LMS algorithm as applied to a simple 2-point FIR. Limiting the number of coefficients to two enabled graphing of the performance surfaces in a three-dimensional space. Adaptive FIRs are employed using large numbers of taps, sometimes in the thousands. It is not possible to visualize the performance surfaces for numbers of taps beyond two, but nonetheless, the LMS algorithm behaves very similarly–that is to say, as a stochastic gradient search algorithm, in which the gradient estimate is noisy. With optimum conditions, typical filters converge in a number of iterations equal to about 10 times the filter length.

The weaknesses uncovered in the 2-Tap FIR are also weaknesses of the general N-Tap LMS adaptive FIR; that is, unreliability with input signals that lack randomness or are of narrow bandwidth. Pure sinusoids, and small combinations of the same can be problematic, and can lead to convergence to incorrect coefficient values.

Reference [1] is a very accessible basic adaptive filtering text; while Reference [2] gives an exhaustive rundown of the LMS adaptive filter and numerous species thereof, along with extensive theoretical discussions, illustrations, and computer exercises.

1.12   REFERENCES

[1] Bernard Widrow and Samuel D. Stearns, *Adaptive Signal Processing*, Prentice-Hall, Englewood Cliffs, New Jersey, 1985.

[2] Ali H. Sayed, *Fundamentals of Adaptive Filtering*, John Wiley & Sons, Hoboken, New Jersey, 2003.

[3] Simon Haykin, *Adaptive Filter Theory*, *Third Edition*, Prentice-Hall, Upper Saddle River, New Jersey, 1996.

1.13   EXERCISES

1. Write the m-code for the script

$$LVxGradientViaCP(FirstValX, StepSize,...$$
$$NoiseAmp, UseWgtingFcn, deltaX, NoIts)$$

which is extensively described and illustrated in the text, and test it with the calls below. For each instance, characterize the performance of the algorithm with respect to speed of convergence and stability. Conditions of stability might be, for example: stable, convergent; stable, but nonconvergent, unstable or divergent. For convergent situations, note where there is overshoot (an "underdamped") condition, very slow convergence (an "overdamped" condition) or immediate convergence (a "critically damped" condition).

    (a) **LVxGradientViaCP(4,0.05,0.01,0,0.0001,50)**
    (b) **LVxGradientViaCP(4,0.03,0.01,0,0.0001,75)**
    (c) **LVxGradientViaCP(4,0.9,0.01,0,0.0001,50)**
    (d) **LVxGradientViaCP(4,1,0.01,0,0.0001,50)**
    (e) **LVxGradientViaCP(4,0.07,0.01,1,0.0001,50)**
    (f) **LVxGradientViaCP(4,0.03,0.01,1,0.0001,20)**
    (g) **LVxGradientViaCP(400,0.03,0.01,1,0.0001,75)**

2. Using the relationships

$$CF[n] = x^2 + 1$$

$$x[n+1] = x[n] - \Delta C$$

$$\Delta C = \mu \left(\partial(CF)/\partial(x[n]) \right) CF[n]$$

with μ = 0.03, write an expression that is a polynomial in x that can be used to solve for an initial value of x (denoted as $x[1]$) that will result in $x[2]$ being equal or very nearly so to the value of x corresponding to the minimum of the performance curve. There should be three values of x, one of which is 0.0. Test your nonzero values of x with the call

LVxGradientViaCP(x,0.03,0,1,delX,6)

where *delX* is given the following values:

    (a) 0.001;
    (b) 0.00001;
    (c) 0.00000001
    (d) 0.0000000001

3. Write the m-code for the script

LVxModelPtswLine(TestM, TestYInt, xTest,...

Mu, yMu2MuRatio, MStart, yIntStart, NoIts)

as described and illustrated in the text, and test it with the given test calls:

    function LVxModelPtswLine(TestM,TestYInt,xTest,Mu,...
    yMu2MuRatio,MStart,yIntStart,NoIts)
    % The line to be modeled is represented in the point-slope form,
    % and thus the test line creation parameters are TestM (the slope),
    % TestYInt (the test Y-intercept) and the vector xTest.
    % Mu is the usual weight for the gradient estimate update

% **term; yMu2MuRatio is a ratio by which to relatively**
% **weight the partial derivative of y-intercept relative to**
% **slope; MStart is the initial estimate of slope to use;**
% **yIntStart is the initial estimate of y-intercept to use,**
% **and NoIts is the number of iterations to perform.**
% **The script creates separate 3-D subplots on one figure**
% **of the performance surface and the coefficient track,**
% **both plots having the same axis limits enabling**
% **easy comparison.**
% **Test calls:**
% **(a) LVxModelPtswLine(2,0,[-10:1:10],0.014,37,-10,8,12)**
% **(b) LVxModelPtswLine(2,0,[-10:1:10],0.005,1,-10,8,40)**
% **(c) LVxModelPtswLine(2,0,[-10:1:10],0.025,1,-10,8,60)**
% **(d) LVxModelPtswLine(2,0,[-10:1:10],0.025,40,-10,8,30)**
% **(e) LVxModelPtswLine(20,-90,[-10:1:10],0.005,10,-10,8,40)**
% **(f) LVxModelPtswLine(2,0,[0:1:10],0.01,50,-10,8,80);**
% **(g) LVxModelPtswLine(1,-2,[0:1:10],0.01,50,-10,8,80);**

The results from using LabVIEW for call (c) above, with separate plots for the performance surface and the coefficient track, are shown in Fig. 1.26.

The result from using MATLAB, with the same arguments as in call (c) above, but with the script written to use a single plot having the coefficient track plotted on top of the performance surface, is shown in Fig. 1.27.

4. Write the m-code for a script

$$LVxModelLineLMS\_MBX(M, B, xVec, Mu, bMu2mMuRat, NoIts)$$

in which the algorithm update equations are Eqs. (1.17)–(1.18) in the text. The input arguments M and B are to be used to generate a set of y-values with corresponding to the x-values $xVec$. The input argument $bMu2mMuRat$ is a number to boost the value of Mu for updating the estimate of b.

Include code within the iterative loop to control Mu to maintain stability, including code to provide separate values of Mu for updating the estimates of m and b. The following code will return the algorithm to a stable condition if Mu is chosen too high, rolling back the coefficients to the previous values and halving Mu if MSE increases.

if Ctr>1
if MSE(Ctr)>1.02*MSE(Ctr-1)
m(Ctr+1) = m(Ctr); b(Ctr+1) = b(Ctr);
Mu = Mu/2
end; end

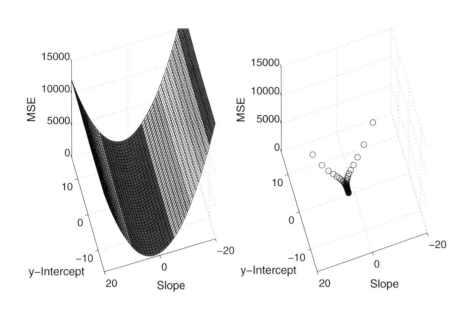

Figure 1.26: (a) Performance Surface; (b) Separate plot of coefficient track along the Performance Surface in (a).

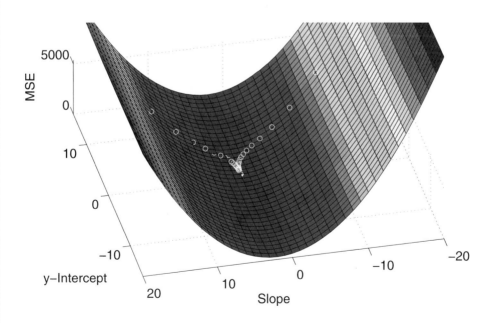

Figure 1.27: Same as the previous figure, except that the coefficient track is plotted onto a plot of the performance surface.

The script should, at the end of each call, create a figure with two subplots, the first of which plots $xVec$ and the corresponding y-values, and a line created using the final estimates of m and b. The second subplot should show MSE v. Iteration.

Test your code with the following calls:

(a) LVxModelLineLMS_MBX(3,-1,[0 1 2 3],1.2, 3,30)
(b) LVxModelLineLMS_MBX(3,-1,[-10:1:10],0.5, 30,5)
(c) LVxModelLineLMS_MBX(3,-1,[0:0.1:2],1.15,3,45)
(d) LVxModelLineLMS_MBX(3,-1,[0:1:2],1.1,2,30)

5. Write the m-code to implement the following script, which creates a two-sample Plant characterized by Plant Coefficients $PC1$ and $PC2$ (as shown in Fig. 1.10) and attempts to model it using a 2-tap LMS adaptive FIR, in accordance with the function specification below. It should create a figure having four subplots which show the test signal, the error signal, and the successive estimates of the plant coefficients, which are given the variable names $c1$ and $c2$. A sample plot, which was created using the call

LVxLMS2TapFIR(4,-3,64,3,[],0.5,0)

is shown in Fig. 1.28.

> **function LVxLMS2TapFIR(PC1,PC2, NoIts,tstDType,CosFrq,...**
> **TwoMu,NoiseAmp)**
> **% PC1 and PC2 are Plant Coefficients, to be modeled by the 2-tap**
> **% FIR; NoIts is the test sequence length in samples**
> **% tstDType selects test signals to pass through the Plant and**
> **% the filter, as follows: 1 = Random Noise; 2 = Unit Step;**
> **% 3 = Nyquist Limit (Fs/2); 4 = Half-Band (Fs/4)**
> **% 5 = cosine of freq CosFrq; 6 = Cos(1Hz)+Cos(Fs/4)+...**
> **% Cos(Fs/2); TwoMu is the update term weight**
> **% Random noise having amplitude NoiseAmp is added to**
> **% the test signal selected by tstDType.**
> **% Test calls:**
> **% LVxLMS2TapFIR(4,-3,64,1,[],0.5,0)**
> **% LVxLMS2TapFIR(4,-3,64,2,[],0.5,0)**
> **% LVxLMS2TapFIR(4,-3,64,3,[],0.5,0)**
> **% LVxLMS2TapFIR(4,-3,64,4,[],0.5,0)**
> **% LVxLMS2TapFIR(4,-3,64,5,[25],0.5,0)**
> **% LVxLMS2TapFIR(4,-3,64,5,[1],0.5,0)**
> **% LVxLMS2TapFIR(4,-3,64,6,[],0.5,0)**
> **% LVxLMS2TapFIR(4,-3,64,0,[],1.15,0)**
> **% LVxLMS2TapFIR(4,-3,7,4,[],1.15,0)**

6. Write the m-code for the script *LVxLMSAdaptFiltMuCheck* as specified below:

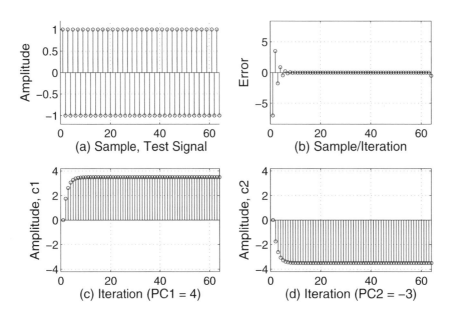

Figure 1.28: (a) Test signal supplied to a two-sample Plant and a 2-Tap LMS Adaptive FIR; (b) Error Signal; (c) Successive estimate plot of first tap coefficient, $c1$; (d) Successive estimate plot of second tap coefficient, $c2$. Note that the test signal is narrow band, and in fact, though the error signal converges to zero, the converged values of $c1$ and $c2$ are incorrect, as expected.

```
function LVxLMSAdaptFiltMuCheck(k,NoTaps,C,SigType,...
Freq,Delay)
% k is the standard deviation of white noise or amplitude of
% a cosine to use as the test signal
% NoTaps is the number of taps to use in the LMS adaptive filter;
% The value for Stepsize is computed as C/(10*P*NoTaps) where
% P = sum(testSig.^2)/length(testSig);
% SigType 0 gives a DC test signal (k*cos(0))
% SigType 1 = random noise, 2 = cosine of frequency Freq
% SigType 3 = Nyquist rate, and 4 = Half-Band Frequency
% The current filter output sample is subtracted from the current
% test signal sample, delayed by Delay samples. Delay must be
% between 1 and NoTaps for the error signal to be able to go
% to zero. Two plots are created, one of the test signal, and
% one of the LMS adaptive FIR error/output signal
% A typical call:
```

% LVxLMSAdaptFiltMuCheck(1,100,1,1,[15],50)
% LVxLMSAdaptFiltMuCheck(1,100,4,1,[15],50)
% LVxLMSAdaptFiltMuCheck(1,100,6,1,[15],50)
% LVxLMSAdaptFiltMuCheck(1,100,8,1,[15],50)
% LVxLMSAdaptFiltMuCheck(1,100,9.5,1,[15],50)
% LVxLMSAdaptFiltMuCheck(1,100,10,1,[15],50)

7. Write the m-code for the script

$$LVxLMSvNLMS(kVec, NoTaps, Delta, Mu, SigType, Freq, Delay)$$

according to the following function specification, and test it with the given calls.

function LVxLMSvNLMS(kVec,NoTaps,Delta,Mu,...
SigType,Freq,Delay)
% kVec is a vector of amplitudes to impose on the test signal as
% a succession of equally spaced amplitude segments over a
% test signal of 15 times the filter length.
% NoTaps is the number of taps to use in the LMS adaptive filter;
% The value for Stepsize is computed as Delta/(P + sm) where
% P = sum(testSig.^2) for the samples of testSig in the filter,
% and this is used in the NLMS algorithm (sm is a small
% number such as 10^(-6). Simultaneously, the test signal is
% processed by an LMS filter of the same length using a
% constant stepsize equal to Mu.
% SigType 0 gives a DC test signal,
% SigType 1 = random noise, 2 = cosine of frequency Freq
% SigType 3 = Nyquist rate, and 4 = Half-Band Frequency
% The current filter output sample is subtracted from the current
% test signal sample, delayed by Delay samples. Delay must
% be between 1 and NoTaps for the error signal to be able to
% go to zero. Five subplots are created, one of the test signal,
% one of the LMS adaptive FIR error/output signal, one of the
% LMS final tap weights, one of the NLMS error signal, and
% one of the final NLMS tap weights.
% Test calls:
% LVxLMSvNLMS([1,5,15,1],100,0.5,0.00005,1,[15],50)
% LVxLMSvNLMS([1,5,25,1],100,0.5,0.00005,1,[15],50)

8. Consider the problem of modeling a set of test points $Pt_i = (x_i, y_i)$ with the polynomial

$$y = a_0x^0 + a_1x^1 + a_2x^2 + ... + a_mx^m \tag{1.33}$$

Since Eq. (1.33) will undergo an iterative process, we designate y and the various coefficients as functions of sample or iteration index n.

$$y[n] = a_0[n]x^0 + a_1[n]x^1 + a_2[n]x^2 + \ldots + a_m[n]x^m \tag{1.34}$$

Equation (1.34) states that the value of y at sample time n, for a certain input x, is equal to the sum of various powers of x, each weighted with a coefficient which is in general a function of iteration n (we seek, of course, to cause all the coefficients, over a number of iterations, to converge to their ideal values, ones which yield the curve which best fits the data points (x_i, y_i)).

We write an expression for total MSE at iteration n, $MSE[n]$, which is the sum of the MSE at each data point:

$$MSE[n] = \frac{1}{N} \sum_{i=1}^{N} (y_i - (\sum_{j=0}^{m} a_j[n]x_i^j))^2 \tag{1.35}$$

Note that $a_j[n]$ and $MSE[n]$ are continuous functions (evaluated at different discrete times or iterations n), and thus may be differentiated. Equation (1.35) may be interpreted as saying that MSE at sample time n is computed by summing, for each test point (x_i, y_i), the square of the difference between the actual test point value y_i and the predicted value of y_i, which is obtained by evaluating the polynomial expression

$$\sum_{j=0}^{m} a_j[n]x_i^j$$

in which x_i is raised in turn to various powers j (0, 1, 2...m), each of which is weighted with the current estimate of the coefficient for that particular power at sample time n, $a_j[n]$.

Using calculus, derive an analytic expression for the partial derivative of MSE with respect to any coefficient a_k and then write an analytic expression for the coefficient update equation to iteratively generate the best estimates of a_k using a true-MSE-based gradient search algorithm, and then test your equations by embodying them in m-code as specified by the function below. Note the additional guidelines given following the function description.

function LVxFitCurveViaLMS(cTest,xTest,NoiseAmp,Mu0,...
UseWts,NoIts)
% cTest is a row vector of coefficients for a polynomial in
% xTest, with the first element in cTest being the coefficient
% for xTest to the zeroth power, the second element of cTest
% is the coefficient for xTest to the first power, and so forth;
% NoiseAmp is the amplitude of noise to add to the
% points((xTest[n],yTest[n]) which are generated by the
% polynomial expression using the coefficients of cTest;
% Mu0 is the initial value of Mu to use in the LMS algorithm;

% UseWts if passed as 1 applies weights to the coefficient
% update equations to speed convergence of the lower power
% coefficients; pass as 0 to not apply any such weights
% NoIts is the maximum number of iterations to perform.
% The script computes the MSE at each iteration, and if it
% increases, the coefficient estimates are rolled back and the
% value of Mu is halved. The brings Mu to an acceptable
% value if it is chosen too large. In order to standardize the
% values of MuO, the initial value of Mu0 is divided by the
% effective power of xTest "in the filter." Additionally, the
% update equation for each coefficient is effectively weighted
% by a factor to cause all coefficients to converge at roughly
% the same rate. The factor is computed as the xTest power
% for the highest power being modeled divided by the
% xTest power for the power whose coefficient is being updated.
%
% Typical calls might be:
%
% LVxFitCurveViaLMS([1, 1, 2],[-5:1:5],0,0.5,1,33)
% LVxFitCurveViaLMS([1, 1, 2],[-5:1:5],0,0.5,0,950)
% LVxFitCurveViaLMS([0, 1, 0, 2],[-5:1:5],0,0.5,1,85)
% LVxFitCurveViaLMS([1, 1, 2],[-1:0.025:1],0,2.75,1,18)
% LVxFitCurveViaLMS([1, 1, -2],[-1:0.025:1],0,4,1,26)
% LVxFitCurveViaLMS([1, 1, -2],[-1:0.025:1],0,1,0,50)

The following code can be used to normalize the stepsize to the signal power of $xTest$ (as described in the function specification above) and to generate the relative $xTest$ signal powers to weight the coefficient update equation to accelerate the convergence of the lower powers, which generally converge much more slowly than the highest power (why?). The computations performed by the code are analogous to computing the signal power in the filter as used in the NLMS algorithm. Here, of course, the signal is the constant vector $xTest$.

```
PwrxVec = 0;
for PwrCtr = 0:1:HighestPower
 PartDerPwr(1,PwrCtr+1) = sum(xTest.^(2*PwrCtr));
 PwrxVec = PwrxVec + PartDerPwr(1,PwrCtr+1);
end
PwrxVec = sum(PwrxVec);
Mu(1) = Mu0/PwrxVec; % normalize Mu[1]
% Compute relative signal power for each polynomial coeff
n = 1:1:HighestPower + 1;
```

if UseWts==1
MuPwrWts(1,n) = ...
PartDerPwr(1,length(PartDerPwr))./PartDerPwr;
else;
MuPwrWts(1,n) = 1;
end

A useful thing in the algorithm is to update Mu each iteration, thus making it a function of discrete time, $Mu[n]$. This can quickly return the algorithm to stability if too large a value of $Mu0$ is chosen. To update $Mu[n]$, compute MSE at each iteration, and if MSE has increased at the current sample, reset (rollback) the coefficient estimates to the previous values, and divide $Mu[n]$ by 2. If MSE is the same or less than for the previous iteration, maintain $Mu[n]$ at the existing value, i.e., let $Mu[N + 1] = Mu[n]$. Note that the stepsize for the coefficient update equation, not including the term $MuPwrWts$, is **2*Mu[n]** to work with the values of $Mu0$ in the test calls given above in the function specification.

CHAPTER 2

Applied Adaptive Filtering

2.1 OVERVIEW

In the previous chapter, we explored a number of basic principles of adaptive processes and developed the LMS algorithm for an FIR arranged in a simple system-modeling problem. In this, we'll investigate several applications of the LMS-based adaptive FIR, including:

1. Active Noise Cancellation
2. System Modeling
3. Echo Cancellation
4. Periodic Signal Removal/Prediction/Adaptive Line Enhancement
5. Interference Cancellation
6. Equalization/Inverse Filtering/Deconvolution
7. Dereverberation

We have briefly seen, in the previous chapter, the filter topology (or arrangement) for system modeling, which is the same as that for Active Noise Cancellation (ANC). The converged coefficients of the adaptive filter that model the Plant in a system-modeling problem necessarily produce a filter output the same as that of the Plant. By inverting this signal and feeding it to a loudspeaker in a duct, noise in a duct can be cancelled. Another common use of active noise cancellation is in headphones being used in a noisy environment. Such headphones can be configured to cancel environmental noise entering the ear canal while permitting desired signals from the headphone to be heard.

With Echo Cancellation, useful in duplex communications systems (such as telephony), for example, we introduce a new topology, which involves two intermixed signals, one of which must be removed from the other. With this topology, we introduce the very useful **Dual-H** method, which greatly aids convergence in such systems.

Periodic Signal Removal and Adaptive Line Enhancement (both accomplished with the same filter topology) find applications in communications systems—a sinusoidal interference signal that persists for a significant period of time can be eliminated from a signal that consists of shorter duration components. This is accomplished by inserting a bulk delay (called a decorrelating delay) between the input signal and the adaptive filter so that the filter output and the signal can only correlate for persistent periodic signals. The LMS filter output itself enhances the periodic signal, thus serving as an Adaptive Line Enhancer (ALE). By subtracting the enhanced filter output from the signal (thus creating the error signal), the periodic component of the signal is eliminated.

With Interference Cancellation, we introduce an important concept that can be used in many different environments. An Interference Cancellation arrangement uses two inputs, a signal reference input and a noise reference input. This arrangement, while discussed in this chapter in the context of audio, also applies to any other kind of signal, such as a received RF signal suffering interference. In that case, a signal antenna and a noise antenna serve as the inputs. As the speed and economic availability of higher speed digital sampling and processing hardware increase, applications of digital signal processing at ever-higher RF frequencies will develop. There are nonetheless many applications in the audio spectrum, such as cancelling vehicular noise entering a communications microphone along with the operator's speech, helping to increase the effectiveness and reliability of audio communication.

Another topic finding applicability in communications systems is inverse filtering, which can be used to perform several equivalent functions, such as equalization or deconvolution. Equalization in communications systems, for example, becomes more necessary as systems become more digital, and maintenance of magnitude and phase responses in communications channels becomes critical to accurate data transmission. A common problem in television reception or FM radio reception is ghosting (in the TV environment) or multipath interference (in the FM environment). These are manifestations of the same thing, multiple copies of the signal arriving at the receiving antenna and causing an echo-like response. Since the echos are essentially attenuated duplicates of the signal, it is possible to determine the echo delay times and magnitudes and remove them. To illustrate the general idea, we conclude the chapter with a problem in which we generate a reverberative audio signal (an audio signal mixed with delayed, attenuated copies of itself) using an IIR, and estimate the parameters of the reverberative signal using adaptive filtering and autocorrelation. From these parameters we are able to deconvolve the audio signal, returning it to its original state prior to entering becoming reverberative.

2.2   ACTIVE NOISE CANCELLATION

The script

$$LVxLMSANCNorm(PlantCoeffVec, k, Mu, freq, DVMult)$$

(see exercises below) demonstrates the principle of cancelling noise in a duct or similar Plant, such as that depicted in Fig. 2.1. This kind of simple arrangement can work when the sound travelling down the duct propagates in the simplest mode, with pressure constant across the cross-section of the duct, only varying along the length of the duct. This is accomplished by making sure the duct's cross-sectional dimensions are small enough compared to the longest sound wavelength present to prevent higher propagation modes.

For our simulation of noise cancellation in a duct, we'll use an adaptive LMS FIR filter having 10 taps, with the Plant (duct) Coefficients specified in a vector of 10 values as *PlantCoeffVector*. The input argument *k* specifies an amount of noise to mix with two sinusoids having respective frequencies determined as harmonics of *freq*; *Mu* is the usual scaling factor, and *DVMult* is a vector of amplitudes by which to scale the test signal, the purpose of which is discussed below.

Example 2.1.   Simulate active noise cancellation in a duct using the script *LVxLMSANCNorm*.

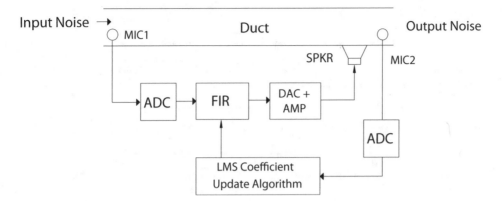

Figure 2.1: In this schematic diagram of a duct noise canceller, MIC1 obtains the noise reference or input for the LMS Adaptive filter (FIR), which is digitized by an analog-to-digital converter (ADC); SPKR (a loudspeaker) emits the negative of the LMS adaptive filter output as a counter-wave to cancel the noise, and MIC2 picks up the residual or error signal, which is also the net duct acoustic output, which is then fed into the LMS coefficient update algorithm block to update the coefficients of FIR.

A typical call might be

LVxLMSANCNorm([0, 0, 1, 0, -0.5, 0.6, 0, 0, -1.2, 0],2,2,3,[1,2,4,8])

which results in Fig. 2.2. It can been seen at plot (a) that all tap coefficients start at a value of zero, and quickly head toward their final values. Tap numbers 3, 5, 6, and 9 can be seen to converge to the same weights found in the Plant's impulse response, *PlantCoeffVec*. Note the random meandering of the coefficient estimates as they slowly, but unerringly, head toward their proper converged values.

The algorithm is an NLMS algorithm (described in the previous chapter), that is to say, it computes the signal power in the filter and uses its reciprocal to scale the coefficient update term; this is what makes the filter stable and immune to the sudden changes in signal amplitude defined by the input argument *DVMult*.

By varying the value of k downward to 0, the deleterious effects on convergence of excessively-low levels of noise in an otherwise-sinusoidal signal can be seen. A typical call that illustrates this, and which results in Fig. 2.3 would be

LVxLMSANCNorm([0,0,1,0,-0.5,0.6,0,0,-1.2,0],0,2,3,[1,2,4,8])

2.3   SYSTEM MODELING

The filter and Plant arrangement in System Modeling is the same as in noise cancellation. Note that in noise cancellation, the noise signal travelled through a system, the Plant, and also through the LMS adaptive filter, which (when properly converged) modeled the Plant's impulse response.

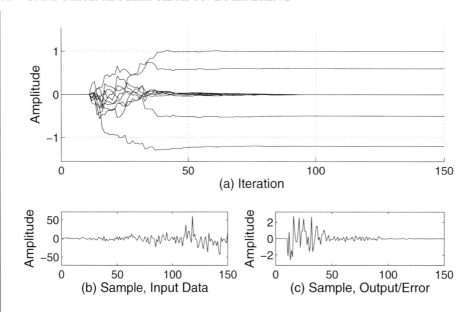

Figure 2.2: (a) All filter Tap Wts versus iteration; (b) Test signal, consisting of noise which has an amplitude profile that increases in steps; (c) The system output (error).

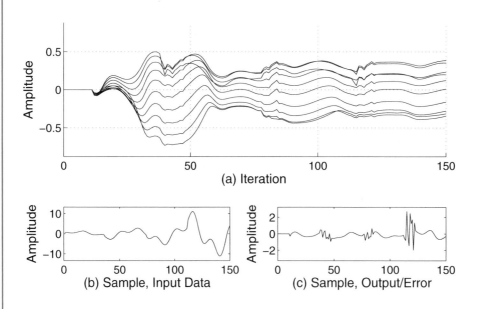

Figure 2.3: (a) Tap Weights versus sample or iteration; (b) Test signal, consisting of sinusoids without noise; (c) The system output, which is the same as the error signal. Note the poor convergence when the input signal is narrowband.

An easy way to obtain any system's impulse response is to send an impulse or other test signal, such as white or pink noise, into the system and record the output. This method is sometimes used in active noise cancellation systems and echo cancellation systems. Often, however, the system cannot be taken offline to determine its impulse response; rather, the system must be modeled (i.e., the impulse response determined) while the system is online, i.e., operating with standard signals rather than test signals.

Example 2.2.   A certain system can be accessed only through its input and output terminals. That is to say, we can inject a test signal into its input terminal, and record the resultant output signal, but we cannot otherwise determine what components the system contains. Assume that the system can be modeled as an IIR filter having a specified number of poles and zeros. Devise a method to obtain its poles and zeros.

We first specify a transfer function consisting of one or more poles and zeros to simulate a system (Plant) to be modeled. We then process a test signal, such as noise or speech, with this Plant to produce a corresponding output sequence. Having these two sequences, we can then model a finite (i.e., truncated) version of the impulse response that produced the output sequence from the input sequence. To obtain the poles and zeros, we use Prony's Method. The script (see exercises below)

$$LVxModelPlant(A, B, LenLMS,$$
$$NoPrZs, NoPrPs, Mu, tSig, NAmp, NoIts)$$

allows you to enter a set of filter coefficients A and B as the numerator and denominator coefficients of the IIR filter that simulates the Plant, the number of LMS FIR filter coefficients $LenLMS$ to use to model the Plant, the number of Prony zeros $NoPrZs$ and poles $NoPrPs$ to use to model the derived LMS coefficients with a pole/zero transfer function, the value of Mu, the type of test signal to pass through the Plant and the LMS filter (0 = White Noise, 1 = DC, 2 = Nyquist, 3 = Half-band, 4 = mixed DC, Half-band, and Nyquist), $NAmp$, the amplitude of white noise to mix with the test signal $tSig$, and the number of iterations to do, $NoIts$. The Prony pole and zero estimates are returned in LabVIEW's Output window.

A typical call might be

LVxModelPlant([1,0,0.81],1,100,3,3,0.5,0,0,1000)

which results in Fig. 2.4 and Prony estimates of b = [0.9986, -0.0008, -0.0013] and a = [1, -0.0003, 0.8099] which round off to b = [1] and a = [1, 0, 0.81].

The same call, modified to call for any test signal other than white noise, generally causes the LMS modeling to fail. Injecting white noise into such an input signal will assist convergence.

Using an LMS algorithm to directly model an IIR-based system is possible, although convergence can be protracted and uncertain; performance surfaces can have localized minima leading

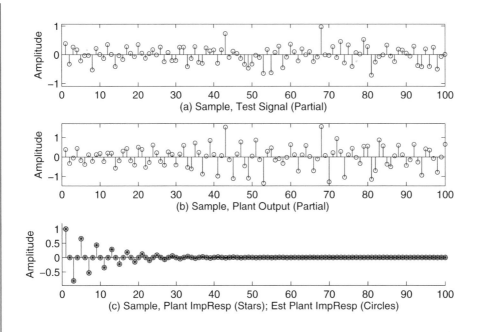

Figure 2.4: (a) Test Signal; (b) Response of Plant to signal at (a); (c) Truncated (actual) impulse response of Plant (stars) and LMS-estimated Plant impulse response (circles).

to incorrect solutions. The method used in this example, although unsophisticated, has the stability and convergence speed of an FIR-based LMS algorithm. Reference [2] supplies several LMS algorithms, including one using an IIR.

2.4   ECHO CANCELLATION

The most common venue for unwanted echo generation is telephone systems, especially loudspeaker telephones. Feedback or howlaround (like the feedback in an auditorium PA system) occurs due to the complete acoustic circuit formed by the loudspeaker and microphone at each end of the conversation. Figure 2.5 shows the basic arrangement.

Referring to Fig. 2.5, we see that a person at the Far End (left side of the Fig.) speaks into microphone M1, the sound is emitted at the Near End by loudspeaker S2, is picked up by microphone M2, and is sent (in the absence of effective echo cancellation at the Near End) down the wire toward the Far End, where it is emitted by loudspeaker S1, and then picked up again by microphone M1, thus forming a complete feedback or echo loop. The adaptive filter ADF2 at the Near End models the acoustic feedback path at the Near End, and trains according to the error signal (the output of the summing junction at the Near End), which is the same as the signal being sent down the wire to the Far End. Since the input to adaptive filter ADF2 is the Far End signal, and only a delayed

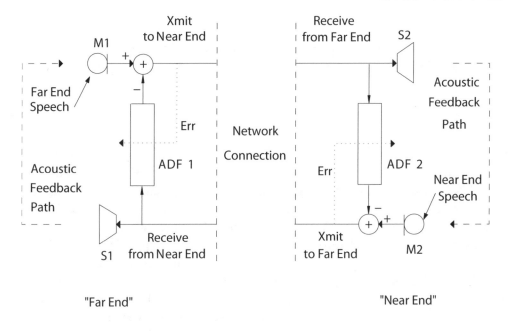

Echo Cancellation Using LMS Adaptive Filters

Figure 2.5: A basic echo cancelling system employed in a telephone system in which both subscriber terminal units are loudspeaker phones, thus leading to the likelihood of acoustic echo, necessitating an echo canceller.

version of it (we are assuming here that the person at the Near End is not speaking) is entering the error junction, the filter converges well.

Problems arise when the person at the Near End speaks–then we have a signal entering the error junction that is not also entering the adaptive filter. The signal from the error junction output thus does not convey an accurate representation of how well the filter is doing in cancelling the Far End signal. As a result, a perfectly converged set of coefficients, which cancel the Far End signal very well, can be caused to severely diverge due to Near End speech.

The patent literature is replete with patents which attempt to solve or regulate the problem caused by Near End speech. The sheer number of such patents suggest that the problem has not been completely solved. Prior to the advent of echo cancellers, a method was used known as **Echo Suppression**. In this method, the levels of Near and Far end signals are compared, and the dominant (usually louder) direction is designated to be the active circuit, and the gain of the nondominant direction is greatly reduced to the point that howlaround no longer exists.

Returning to the problem of simulating an echo cancellation system, the scripts

$$LVxLMSAdaptFiltEcho(k, Mu, DHorSH, MuteNrEnd, ...$$
$$DblTkUpD, SampsVOX, VOXThresh)$$

and

$$LVxLMSAdptFiltEchoShort(k, Mu, DHorSH, MuteNrEnd, PrPC)$$

allow you to experiment with echo cancellation. For the first mentioned script, two audio files, *drwatsonSR8K.wav*, and *whoknowsSR8K.wav* are used as the Near End and Far End signals, respectively. The second script is suitable when script execution speed needs to be improved; it uses two bandwidth-reduced audio files, *drwatsonSR4K.wav*, and *whoknowsSR4K.wav*, and does not compute or archive certain functions. The description of parameters below applies to the first mentioned script, although the first four input parameters are identical for both scripts. A more complete description of the second mentioned script is given in the exercises below.

Playing the unprocessed and processed signals for comparison can be programmed in the script to take place automatically, but for repeat playing, the directions in the exercises below suggest that the relevant variables representing the unprocessed and processed sounds be made global in the script. Then, by declaring them global on the Command line, it is a simple matter to play them using the function *sound*. To convey to the user the appropriate variable names and the sample rate, these are programmed in the script to be printed in the Command Window in MATLAB or Output Window in LabVIEW. For example, after making the call

LVxLMSAdptFiltEchoShort(0.01,0.2,1,1,50)

in LabVIEW, the following is printed in the Output Window:

global sound output variable is EchoErr & sample rate is 8000

To listen to the processed audio at will, enter the following lines, pressing Return after each line:

global EchoErr
sound(EchoErr,8000)

The method for playing audio signals just described also holds true for the other scripts in this chapter that produce audio outputs. When using MATLAB, it is possible to create pushbuttons on the GUI with callback routines to play the audio files on demand (this method is also implemented using global variables).

Both scripts simulate the acoustic path at the Near End as a single delay with unity gain. In the figures below, the plots of coefficients show, under ideal conditions, one coefficient converging to a value of 1, and the nine other coefficients converging to the value 0. As we'll see, however, conditions are not always ideal and often the coefficients remain misadjusted. Prior to conducting a systematic exploration of why this is and what can be done about it, we'll define the input arguments of the script *LVxLMSAdaptFiltEcho*:

- The argument *k* allows you to choose an amount of white noise to mix with the Far End signal to assist the LMS adaptive filter in converging to a set of coefficients which models the acoustic path between S2 and M2 as shown in Fig. 2.5.

- The argument *Mu* is the usual update term weighting constant for the LMS algorithm.

- The argument *DHorSH* tells the script whether or not to use a single LMS adaptive filter (*Single-H*), or to use the *Dual-H* mode, in which an on-line filter and an off-line filter are used, and a figure-of-merit called ERLE is computed to determine how effectively the echo signal is being removed from the error signal, which is the signal being returned to the Far End from the Near End, along with Near End speech. A detailed discussion of the difference between the Single-H and Dual-H modes, along with a mathematical definition of ERLE, is found below.

- *MuteNrEnd* can take on the values 0, 1, or 2; if specified as 0, the Near End signal is emitted continuously into microphone M2. If *MuteNrEnd* is specified as *1*, the Near End signal is muted for a brief period at the beginning of echo cancellation, thus giving the Far End signal a brief period without Near End speech in which to try to converge. If *MuteNrEnd* is specified as 2, the Near End signal is muted once about one-third of the way through. This muting is meant to simulate near silence from the Near End speaker (person); such silence assists the LMS adaptive filter in converging to a set of coefficients which model the acoustic path between the loudspeaker S2 and microphone M2 at the Near End.

- The argument *DblTkUpD*, if passed as 1, prevents coefficient update in Single-H mode whenever Near End speech is detected. If passed as 0, coefficient update is permitted even when Near End speech is present.

- The remaining two arguments, *SampsVOX*, and *VOXThresh*, are two parameters you can manipulate to specify how Near End Speech is detected. The first argument, *SampsVOX*, tells how many samples of the Near End speech signal to use in computing the RMS value, which is then compared to the second argument, *VOXThresh*, which is a threshold below which it is judged that Near End speech is not present.

2.4.1 SINGLE-H

The *Single-H* mode employs a single LMS adaptive filter to cancel a Far End acoustic echo that is "contaminated" with Near End speech. Such an arrangement performs poorly unless coefficient update is stopped whenever Near End speech is detected. The script *LVxLMSAdaptFiltEcho* allows you to specify *Single-H* and whether or not to allow coefficient update during Near End speech (sometimes called "doubletalk," since both Near End and Far End are making sound).

Example 2.3.   Demonstrate adaptive echo-cancelling using the Single-H mode.

Suitable calls to use the *Single-H* mode with coefficient update during doubletalk would be

LVxLMSAdaptFiltEcho(0.02,0.2,0,0,0,50,0.02)

LVxLMSAdptFiltEchoShort(0.02,0.2,0,0,50)

Figure 2.6 shows the result of making the first call (the second script call plots only the output/error signal) for *Single-H* echo cancellation with update during doubletalk. Note that in plot (a) the coefficient values in the adaptive filter, which should converge to essentially constant values, are overladen with noise. Note that the Single-H mode does not use the parameter ERLE, but ERLE is computed and displayed for comparison with the Dual-H mode, in which ERLE increases monotonically as better coefficient estimates occur.

Example 2.4.   Demonstrate adaptive echo-cancellation using the Single-H mode, but freeze coefficient update during doubletalk.

Freezing coefficient update during doubletalk results in considerable improvement, as can be seen in Fig. 2.7, which resulted from the script call

LVxLMSAdaptFiltEcho(0.02,0.2,0,0,1,50,0.02)

Here we see that the filter coefficients, shown in plot (a), do assume reasonably constant values, but with a much decreased level of noise, since coefficient update during Near End speech was inhibited, as shown in plot (d), where a value of zero at a given sample or iteration means that coefficient update was inhibited for that sample/iteration.

2.4.2   DUAL-H

The Dual-H mode is a simple way to overcome the difficulties caused by Near End speech. In the Dual-H mode, a main filter or on-line filter, which actually generates the counter-echo signal which is subtracted from the Near End's microphone output (M2 in Fig. 2.5), has its coefficients updated only when a second, off-line filter, which adapts continuously (i.e., computes a new set of coefficients for each signal sample), generates a set of filter coefficients which result in a better measure of merit, such as ERLE (Echo Return Loss Enhancement), which is defined here and in the script *LVxLMSAdaptFiltEcho* (and others that follow in this chapter) as the ratio of Far End power (or energy) to Near End (output of LMS error junction) power (or energy). These values can, for better effect, be averaged over a number of samples (in the scripts mentioned above, the

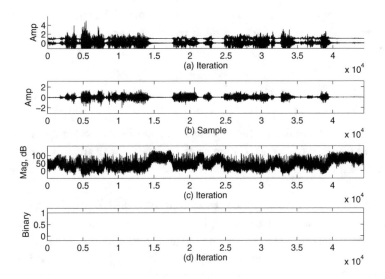

Figure 2.6: (a) All of the adaptive filter coefficients versus sample or iteration; (b) Error signal; (c) ERLE, a figure of merit representing how much of the Far End signal has been removed from the signal; (d) Function "Coefficient Update Permitted?" A value of 0 means that coefficient update was inhibited due to the presence of Near End speech, while a value of 1 means coefficient update was allowed to proceed at the particular iteration or sample number.

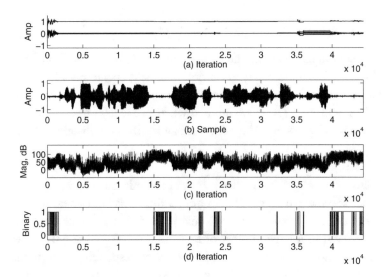

Figure 2.7: (a) All of the adaptive filter coefficients versus sample or iteration; (b) Error signal; (c) ERLE. Note that, due to doubletalk, there was relatively little coefficient update during the course of the conversation, but the performance was improved considerably over the constant coefficient update scenario; (d) Function "Coefficient Update Permitted?"

number chosen is equal to the filter length, but this need not be the number). Thus, ERLE might be defined as the ratio of the average signal power input to the filter to the average error power for the corresponding samples, i.e.,

$$ERLE[n] = (\sum_{i=n-N+1}^{n} S[i]^2)/((\sum_{i=n-N+1}^{n} E[i]^2) + \epsilon) \qquad (2.1)$$

where n is the sample index, N is the chosen number of samples to use in the computation, S represents the signal (the Far End signal in this case), E represents the error signal, and ϵ is a small number to prevent division by zero if the error signal is identically zero over the subject interval of samples.

The current best set of filter coefficients and a corresponding best ERLE are stored and for each new sample, the off-line filter is updated according to the LMS algorithm, and an echo prediction signal (i.e., LMS adaptive filter output) is generated and subtracted from the current Near End microphone output to generate a Test Error signal. A value for ERLE is generated based on the Test Error, and if this Test ERLE is improved over the current best ERLE, the off-line or test coefficients are stored into the on-line filter and also as the current best coefficients. The best ERLE is replaced by the newly computed Test ERLE.

Example 2.5.   Demonstrate adaptive echo-cancelling using the Dual-H mode.

Figure 2.8 shows the results from the script call

LVxLMSAdaptFiltEcho(0.02,0.2,1,1,[],[],[])

in which the last three arguments are immaterial since the Dual-H mode has been specified by the third argument. In the Dual-H system, it is unnecessary to restrict coefficient update since ERLE is used to judge whether or not coefficients in the on-line filter should be updated, and hence the input parameters *DblTkUpD*, *SampsVOX*, and *VOXThresh* are ignored when in Dual-H mode, and the function "Coefficient Update Permitted" is assigned the value 1 for all iterations of the algorithm.

2.4.3   SPARSE COMPUTATION

The plant delay in echo cancellers often runs into hundreds of milliseconds and hence, at, say, a sample rate of 8 kHz, an FIR that covered every possible tap from a time delay of one sample up to and somewhat beyond the length of the impulse response could potentially be several thousand taps long. Often in echo cancellers, out of a large number of taps, however, there are only a few taps having significant amplitude. These taps are, of course, located at corresponding echo times represented by the impulse response of the echo-causing system (a room, for example, in the case of a speakerphone). A common technique to greatly reduce the needed computation is to estimate the locations of the taps having significant amplitudes, and to restrict tap computation to relatively short runs of contiguous taps around each significant tap location. This technique also helps to ensure

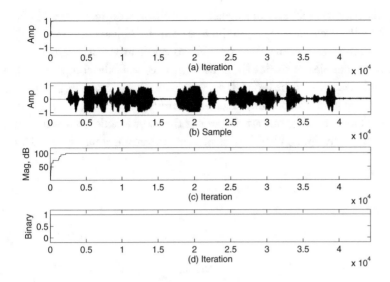

Figure 2.8: (a) All of the adaptive filter coefficients versus iteration; (b) Error signal; (c) Current Best Value of ERLE. Note that the brief introductory portion of the signal was muted, allowing the system to quickly converge to an excellent set of coefficients, which were not improved upon as ERLE remained constant after the initial convergence; (d) Function "Coefficient Update Permitted?"

proper convergence since all noncomputed taps are inherently valued at zero. The selection of taps to be computed, and thus also those not to be computed, is itself a filtering process. One method to estimate the relevant taps is to cross correlate the Far End signal with the signal coming from the Near End microphone, which contains an echoed version of the Far End signal plus the Near End signal.

As sparse computation largely arises from practical and empirical knowledge and practice, an excellent source for information is the website www.uspto.gov, which allows you to do word searches for relevant U.S. Patents, which contain much information not found in other sources. A number of patents relating to sparse computation, which may be viewed or downloaded from www.uspto.gov, are cited below. You may also conduct your own searches. The best (most focused and pertinent) results are usually obtained when a very specific search term is used, such as "pnlms," "nlms," "lms," etc. Advanced searches can also be performed using multiple terms, such as "sparse" and "echo," for example.

2.5   PERIODIC COMPONENT ELIMINATION OR ENHANCEMENT

Figure 2.9 shows the basic arrangement for predictive filtering. Since the input to the filter is delayed, it is not possible for it to adapt to the input signal in time to cancel the input signal at the error summing junction. Of course, this is only true for random signals which change constantly. A periodic signal, such as a sine wave, looks alike cycle after cycle, so the filter can still adjust and cancel it in the error junction's output. The implication is that it is possible to select or enhance a periodic component with the adaptive filter. The filter's output yields the enhanced periodic component(s), and the error junction output yields a version of the input signal with periodic components removed. This is useful for eliminating periodic interference/noise, such as heterodynes (whistles or tones) in a communications receiver or the like. The filter output per se (not the error junction output) is an enhanced version of any sinusoidal input signal. Used in this manner, the arrangement is called an **Adaptive Line Enhancer**.

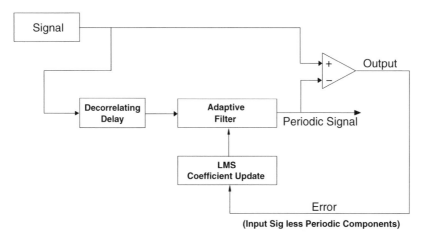

Periodic Component Cancellation

Figure 2.9: A basic arrangement which uses an LMS Adaptive filter to eliminate only periodic components from a signal.

The scripts

$$LVxLMSAdaptFiltDeCorr(k, Mu, freq, NoTaps, DeCorrDelay)$$

and

$$MLx\_LMSAdaptFiltDeCorr(k, Mu, freq, NoTaps, DeCorrDelay)$$

(see exercises below) demonstrate elimination of periodic signals mixed with white noise, leaving mainly the white noise. The first input argument specifies the amplitude of white noise, the second argument is Mu, and the third argument, $freq$, is used to generate two sine waves having, respectively, frequencies of *freq* and 3*freq. The sampling rate is 1024 for this script, so the maximum number you can use for the third argument is about 170 (without aliasing). The fourth argument allows you to specify the number of taps used in the filter, up to 100, and the fifth argument allows you to specify the number of samples of decorrelating delay. In the MATLAB-suitable script *MLx_LMSAdaptFiltDeCorr*, the various signals may be heard by pressing buttons on the GUI created by the script. In the LabVIEW-suitable script, the sounds play automatically, and the sound files are global variables that can be played again by declaring the variables global on the Command line and then giving suitable Commands. An example of this is given immediately below.

Example 2.6. Using the script *LVxLMSAdaptFiltDeCorr*, demonstrate periodic component elimination. Use a decorrelating delay of 2 samples.

Suitable calls would be

$$\textbf{LVxLMSAdaptFiltDeCorr(0.2,1,100,10,2)}$$

or

$$\textbf{MLxLMSAdaptFiltDeCorr(0.2,1,100,10,2)}$$

which result in Figs. 2.10 and 2.11. Much of the amplitude in the input signal is from the two periodic components. After the filter converges (very quickly in this case), only the residual white noise is left; this is also evident from the respective DFTs, plotted to the right.

After making the script call above, the following information will be printed in the Command Window:

Comment =
global sound variable names are LMSDeCorDataVec, Err, and FiltOut
Comment =
global sound variable sample rate is 8000

This information can be used to listen to the audio files on demand, as described earlier in the chapter. Note that the sample rate in the script is 1024 Hz, but LabVIEW only accepts standard sample rates such as 8000, 11025, 22050, and 44100. Thus, the sounds produced by LabVIEW will be higher in pitch by the factor 8000/1024 and shorter in duration by the factor 1024/8000. When writing and running the MATLAB-suited script, you can use the actual sample rate of 1024 and use pushbuttons on the GUI or use the global variable method described above.

Example 2.7. Show that when the decorrelating delay is zero samples, the script *LVxLMSAdapt-FiltDeCorr* will no longer decorrelate the filter and algorithm update inputs, and hence the entire input signal will be cancelled.

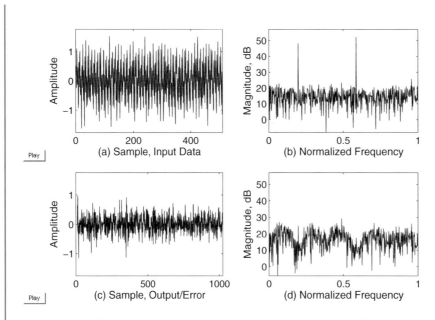

Figure 2.10: (a) Test signal, consisting of periodic components of high magnitude and noise of much lower magnitude; (b) Spectrum of the test signal; (c) Output/Error; (d) Spectrum of Output/Error, showing only noise.

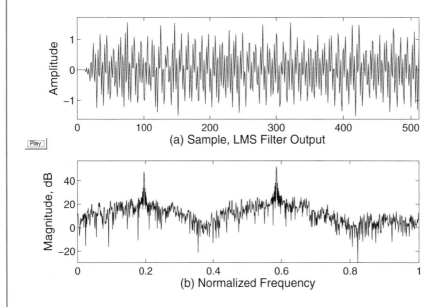

Figure 2.11: (a) The output of the LMS adaptive FIR, showing periodic signals enhanced by the attenuation of noise; (b) Spectrum of signal in (a). Note the attenuation of noise, in contrast to the unattenuated noise-and-signal spectrum shown in Figure 2.10, plot (b).

The call

LVxLMSAdaptFiltDeCorr(0.2,1,100,10,0)

which specifies the decorrelating delay as 0, results in the filter being able to cancel both periodic and nonperiodic components, as shown in Fig. 2.12 (this call essentially configures the filter as an Active Noise Canceller).

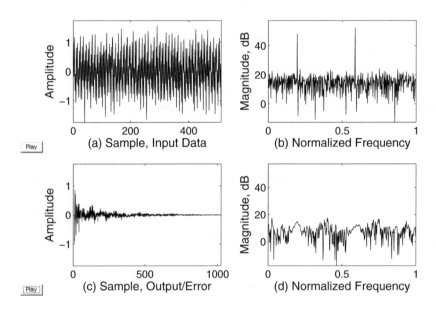

Figure 2.12: (a) Test signal, consisting of periodic components of high magnitude and noise of much lower magnitude; (b) Spectrum of the test signal; (c) Output/Error; (d) Spectrum of Output/Error, showing attenuation of both periodic and aperiodic components.

Two scripts, *LVxLMSDeCorrHeterodyne* and *LVxLMSDeCorrHetDualH*, are developed in the exercises below to explore periodic component elimination in more detail.

2.6 INTERFERENCE CANCELLATION

Another use for adaptive filtering is to cancel interference from a channel signal which is a mixture of some desired signal and an interference signal. A reference signal is obtained for the interference, and it is passed through the adaptive filter and the result subtracted from the channel signal. In the general case, the reference signal also contains a component of the desired signal.

Signal Plus Noise and Noise Plus Signal Channels

Figure 2.13 shows a basic arrangement for interference cancellation. Interference cancellation is implemented by using a separate input which serves as a reference signal for the noise (interference) component. The reference signal, ideally, consists only of the noise signal and is unpolluted with the desired signal component. In many, if not most practical cases, this will not be true, but effective cancellation is usually possible. The general case is shown here, where some of the desired signal mixes with the noise reference, and of course some of the noise mixes with the desired signal (if it didn't, there would be no interference needing to be cancelled). This arrangement works, delay-wise, when the desired signal source is close to M1 (i.e., Path 1 is very short) and the noise source is close to M2 (i.e., Path 3 is very short), since in this situation the delay of the LMS adaptive filter can effectively model the delay from the noise source to M1, which is the delay of Path 4. On the other hand, the delay from the desired signal source to M2 (via Path 2) is such that the filter cannot correlate the desired signal component in the error signal with the desired signal component in the filter. Hence, the net result is that the noise signal can be cancelled, and the desired signal cannot be cancelled. The other thing to note is that the desired signal distorts the error signal, just as the near end signal does in an echo canceller. As a result, a Dual-H arrangement or a Single-H arrangement with no update during desired signal speech must be used.

Referring to Fig. 2.13, the problem of cancelling the Noise Signal component from the Output signal can be seen as a problem of linear combination. Here we are making the assumption, for purposes of ease of explanation, that all delays associated with Paths 1 through 4 are simple delays (i.e., no echoes are involved) so that each path delay can be simply characterized as z^{-N} where N represents the path delay as a number of multiples N of the sample period.

Using z-transform notation we get

$$Output(z) = G_1 S(z) z^{-\Delta 1} + G_4 N(z) z^{-\Delta 4} ...$$

$$-(G_3 N(z) z^{-\Delta 3} + G_2 S(z) z^{-\Delta 2}) LMS(z)$$

where G_1, for example, means the gain over Path 1 ($G_i = 1$ means no attenuation, $G_i = 0$ means complete attenuation), $Output(z)$ is the z-transform of the output signal, $S(z)$ means the z-transform of the Desired Signal, $N(z)$ means the z-transform of the noise, $z^{-\Delta 1}$ represents the delay of Path 1, etc., and $LMS(z)$ means the z-transform of the LMS adaptive filter.

When the filter has converged, the Noise component that is mixed with the Desired Signal is cancelled in the output, and hence it is possible to say that

$$G_4 N(z) z^{-\Delta 4} = G_3 N(z) z^{-\Delta 3} LMS(z)$$

which simplifies to

$$LMS(z) = G_4 N(z) z^{-\Delta 4} / (G_3 N(z) z^{-\Delta 3})$$

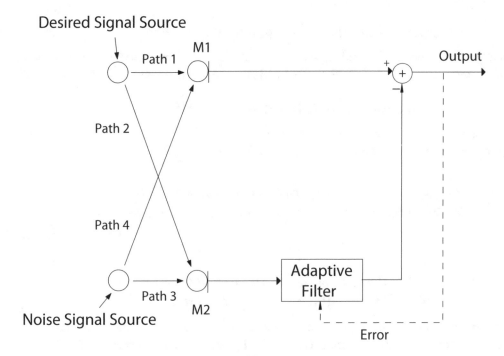

Interference Cancellation

Figure 2.13: A basic interference-cancelling arrangement.

and finally

$$LMS(z) = \frac{G_4}{G_3} z^{-\Delta 4 + \Delta 3}$$

The important thing to note is that the LMS adaptive filter imparts time *delay* only, not time advancement, which means that the quantity $(-\Delta 4 + \Delta 3)$ must remain negative and have a value (in terms of number of samples) that lies within the capability of the filter.

For example, if $\Delta 4 = 8$ samples, and $\Delta 3 = 1$ sample, then the quantity $-\Delta 4 + \Delta 3 = -7$ which means that the LMS adaptive filter must be able to impart a delay of 7 samples. If $\Delta 3$ is greater than $\Delta 4$, of course, the filter would be expected to advance signals in time, which it cannot do (if $\Delta 3 = \Delta 4$, the filter must have a zero-delay tap).

At proper convergence, the Noise Signal is completely cancelled and the output becomes

$$Output(z) = G_1 S(z) z^{-\Delta 1} - G_2 S(z) z^{-\Delta 2} \frac{G_4}{G_3} z^{-\Delta 4 + \Delta 3}$$

which simplifies to

$$Output(z) = S(z)(G_1 z^{-\Delta P1} - (G_2 G_4 / G_3) z^{-\Delta 2 - \Delta 4 + \Delta 3})$$

What this says is that the Desired Signal is effectively filtered by the interference cancellation process. For example, assuming that $G_1 = G_3 = 1$ and $\Delta 1 = \Delta 3 = 0$, we get

$$Output(z)/S(z) = (1 - G_2 G_4 z^{-\Delta P2 - \Delta P4})$$

which is the transfer function of a modified comb filter.

The scripts (see exercises below)

LVxLMSInterferCancel(k,Mu,DHorSH,MuteDesSig,DblTkUpD,
NoSampsVOX,VOXThresh,GP2,GP4,DP1,DP2,DP3,DP4,NumTaps)

and

LVxLMSInterferCancShort(k,Mu,DHorSH,MuteDesSig,
GP2,GP4,DP1,DP2,DP3,DP4,NumTaps,PrPC)

offer the opportunity to experiment with interference cancellation. A large number of parameters are provided to allow ample experimentation. $GP2$ and $GP4$ have the same meaning as G_2, G_4, etc., used above, and $D1$ means the same thing as $\Delta 1$, etc. The first seven input arguments of the first mentioned script above are the same as the first seven arguments of the script *LVxLMSAdaptFiltEcho*.

For purposes of improving computation speed, the second script, *LVxLMSInterferCancShort*, uses bandwidth reduced audio files ('drwatsonSR4k.wav' and 'whoknowsSR4K.wav') and does not deal with doubletalk or archive parameters such as the figure of merit (IRLE or Interference Return Loss Enhancement). The input argument $PrPC$ is the percentage of the audio files to process when running the script (30% to 50% will usually suffice to demonstrate the required results). More detail on the latter script, *LVxLMSInterferCancShort* will be found in the exercises below, where the student is guided to create the m-code for the script.

The argument, *NumTaps*, found in both scripts, allows you to specify the number of taps used by the adaptive filter. For proper cancellation, it should be equal to at least *DP4 - DP3*. For example, if *DP4* = 42, and *DP3* = 7, *NoTaps* should be 35 or greater.

For purposes of reducing the number of required input arguments, the gains of paths P1 and P3 are assigned the value 1, with realistic values of G_2 and G_4 being fractions of 1; this corresponds approximately to the situation in which each microphone is very close to its respective sound source, which results in the highest signal-to-noise ratio for M1, and the highest noise-to-signal ratio for M2, which would result in the smallest amount of comb filtering of the desired signal in the output.

Example 2.8.   Simulate interference cancellation using the script *LVxLMSInterferCancel*. Make a call that shows interference cancellation with little comb filtering, and make a second call that shows pronounced comb filtering.

For the first task, we make the following call, which has relatively low gain (1/16) for paths 2 and 4, and which specifies use of the Dual-H mode with immediate muting to speed convergence, and which results in Fig. 2.14.

LVxLMSInterferCancel(0.02,0.3,1,1,0,50,0.03,1/16,1/16,6,24,6,24,18)

A call requiring less computation, and which plots only the filtered output signal (see exercises below for further details) is

LVxLMSInterferCancShort(0.02,0.3,1,1,1/16,1/16,6,24,6,24,18,30)

The spectrum of the output signal is shown in Fig. 2.15 in plot (a); this may be contrasted with the spectrum shown in plot (b), which arises from the call

LVxLMSInterferCancel(0.02,0.3,1,1,0,50,0.03,1,1,1,6,1,6,18)

in which the paths P2 and P4 are given gains of 1.0 (i.e., no attenuation). A comb-filtering effect is clearly visible in plot (b), where the cross-paths P2 and P4, like paths P1 and P3, have gains of 1.0. In realistic situations, where the gains of paths P2 and P4 are small compared to those of P1 and P3, the comb-filtering effect is greatly attenuated, as is evident in plot (a) of Fig. 2.15.

Reference [1] gives a basic description of the various LMS filtering topologies discussed in this chapter.

2.7   EQUALIZATION/DECONVOLUTION

It is possible to use an adaptive filter to equalize, or undo the result of convolution; that is to say, a signal that has passed through a Plant (being, perhaps, an amplification channel having a filtering characteristic such as lowpass or bandpass) will have its characteristics changed since it will have been convolved with the Plant's impulse response. A filter that can undo the effects of convolution might be called a deconvolution filter, or an equalization filter, or simply an inverse filter. For simplicity, we'll only use one of these terms in the discussion below, but all of them apply.

The basic arrangement or topology for channel equalization is as shown in Fig. 2.16.

The script

$$LVxChannEqualiz(ChImp, Mu, tSig, lenLMS)$$

allows experimentation with channel equalization using an LMS algorithm in the topology shown in Fig. 2.16. The parameter $ChImp$ is the impulse response of the Channel, Mu has the usual meaning in LMS adaptive filtering, $tSig$ is an integer that selects the type of test signal to pass through the Channel/LMS Filter serial combination to attempt to train the filter to the inverse of the Channel Impulse Response, with 0 yielding random noise, 1 yielding a repeated unit impulse sequence of length equal to the LMS filter, with its phase alternated every repetition, 2 yielding a repeated unit impulse sequence equal in length to the adaptive filter, and 3 yielding a sinusoidal sequence of half-band frequency. The length of the adaptive filter is specified by $lenLMS$.

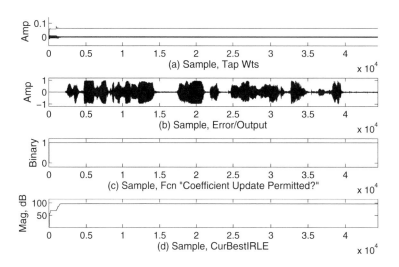

Figure 2.14: (a) All of the adaptive filter coefficients versus sample or iteration, showing good convergence and stability; (b) Output/Error signal; (c) Function "Coefficient Update Permitted?"; (d) Current Best value of Interference Return Loss Enhancement (IRLE) during the course of algorithm execution.

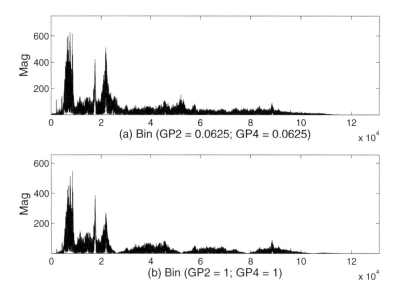

Figure 2.15: (a) Typical output spectrum of signal after noise signal is subtracted, with the condition that significant attenuation exists in the crosspaths P2 and P4; (b) Same as (a), except that there was no attenuation in the crosspaths, leading to significant comb filtering effects in the output signal.

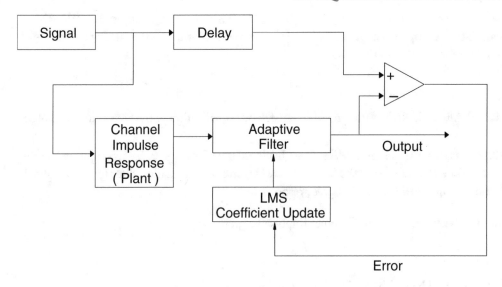

Channel Equalization / Inverse Filtering / Deconvolution

Figure 2.16: A basic arrangement for Channel Equalization.

An important thing to note in the script is that the Channel Impulse Response is always modeled as an FIR, and the adaptive filter is also an FIR. For mathematically perfect inverse filtering, the transfer functions of the two should be the reciprocals of one another. Since both are FIRs, this cannot be. However, an LMS FIR filter of sufficient length can converge to a truncated version of an infinite impulse response.

Example 2.9.   A Channel has z-transform $1 - 0.7z^{-1}$. Determine what length of LMS FIR filter should be used if we arbitrarily require that the truncated inverse include all samples having magnitude of 0.01 or greater.

The reciprocal of the Channel's z-transform is $1/(1 - 0.7z^{-1})$ (ROC: $|z| > 0.7$), and the corresponding impulse response is $[1, 0.7, 0.49, ...0.7^n] = (0.7)^n u[n]$. This is an exponentially decaying sequence, and we therefore need the smallest value of n such that

$$0.7^n \leq 0.01$$

We compute $\log(0.01) = n \log(0.7)$ from which $n = 12.91$ which we round to 13. Thus, an LMS FIR length of 13 or more should be adequate under this criterion.

Example 2.10.   Determine if the impulse response of an IIR, truncated to a length-17 FIR, can serve as an adequate inverse for a Channel having its impulse response equal to $[1, 1]$.

The true inverse system has $H(z) = 1/(1 + z^{-1})$. This is a marginally stable IIR with a nondecaying impulse response equal to $(-1)^n u[n]$. The z-transform of the converged, length-17 Channel/LMS Filter system in this case is

$$(1 + z^{-1})(1 - z^{-1} + z^{-2} - \ldots + z^{-16}) = 1 + z^{-17}$$

which is a comb filter, which does not provide a satisfactory equalization of the net channel response.

Example 2.11.   For a Channel Impulse Response having z-transform $1 + 0.7z^{-1}$, compute a length-17 truncation of the IIR that forms the (approximate) inverse or equalization filter for the Channel. Compute the net equalized Channel impulse response.

Since the z-transform of the Channel Impulse Response is

$$H(z) = 1 + 0.7z^{-1}$$

the true inverse would be an IIR having the z-transform

$$H(z) = \frac{1}{1 + 0.7z^{-1}}$$

which equates to a decaying Nyquist rate impulse response with values $[1,-0.7,0.49,\ldots(-0.7)^n]$ where n starts with 0.

The z-transform of the net Channel/LMS Filter is

$$(1 + 0.7z^{-1})(1 - 0.7z^{-1} + 0.49z^{-2} - \ldots + (-0.7)^{16}z^{-16}) = 1 + (0.7)^{17}z^{-17}$$

which simplifies to $1 + 0.0023z^{-17}$; this is a severely modified comb filter with a miniscule delay term having essentially no effect, i.e., the inverse approximation is a very good one as the net combination approaches having a unity transfer function.

The result from making the call

LVxChannEqualiz([1,0.7],2.15,2,17)

(in which Mu is 2.15 and the test signal comprises multiple repetitions of a unit impulse sequence having the same length as the filter, 17) is shown in Fig. 2.17.

Example 2.12.   Demonstrate Channel Equalization using the script *LVxChannEqualiz*, with a channel impulse response of $[1, 0, 1]$.

The Channel Impulse Response $[1,0,1]$ is a comb filter with precisely one null at the half-band frequency, making it a simple notch filter at that frequency. The IIR which is the perfect inverse

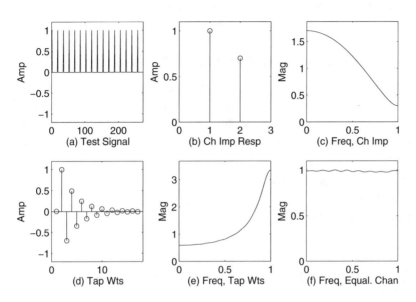

Figure 2.17: (a) The test signal; (b) Channel impulse response; (c) Frequency response of the Channel; (d) Converged tap weights of inverse filter; (e) Frequency response of tap weights shown in (d); (f) Frequency response of net equalized channel.

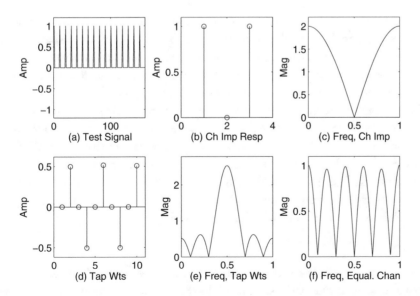

Figure 2.18: (a) The test signal; (b) Channel impulse response; (c) Frequency response of the Channel; (d) Converged tap weights of inverse filter; (e) Frequency response of tap weights shown in (d); (f) Frequency response of equalized channel.

filter has an infinite, nondecaying impulse response forming a sinusoid at the half-band frequency (i.e., four samples per cycle), such as [1,0,-1,0,1,0,...]. The call

$$\textbf{LVxChannEqualiz([1,0,1],2.4,2,10)}$$

which specifies $ChIMp$ = [1,0,1], Mu = 2.4, $tSig$ = repetitive unit impulse sequence of length 10, and $lenLMS$ = 10, results in Fig. 2.18, which shows that the tap weights have converged to a half-band-frequency sinusoid, but the resultant equalized channel response is that of a comb filter. This can be seen by convolving the Channel's impulse response and the LMS filter's impulse response, i.e.,

> **y = conv([1,0,1],[0,1,0,-1,0,1,0,-1,0,1])**
> **% y = [1,0,0,0,0,0,0,0,0,0,1]**
> **fr = abs(fft(y,1024)); plot(fr(1,1:513))**

which results in Fig. 2.19.

2.8   DECONVOLUTION OF A REVERBERATIVE SIGNAL

An occasionally-occurring problem in audio processing is the presence of echo or reverberation, caused by a highly reflective acoustic environment. While small to moderate amounts of reverberation are often imparted to signals deliberately to make them resemble signals in a reverberative room or hall, large amounts of unwanted or unintended reverberation often make an audio signal difficult to comprehend, in the case of speech, or unpleasant in the case of music.

In the previous section we used a topology for deconvolution that assumed that the unconvolved (i.e., un-reverberated) signal is available as a reference. In this section, we will attempt to dereverberate (deconvolve) an audio signal for which we do not possess the pre-reverberative signal.

In theory, if we can estimate the transfer function between the sound source and the microphone (or listener), we can undo all or part of the reverberative (or echo) process. For our study, we'll use a simple type of reverberation created using an IIR with a single stage of feedback having a long delay and real scaling factor with magnitude less than one. The echoes in such a signal are evenly spaced and decay at a uniform rate, making estimation of the parameters relatively easy. We'll use the estimated parameters of the IIR (the delay in samples and the gain of the feedback stage) and form a deconvolution or inverse filter to remove the echo from the original signal. We'll also take a look at the spectral effect of reverberation on the signal. To do all of this, we'll call on knowledge of IIRs, the z-Transform, Autocorrelation, and LMS Adaptive filtering.

2.8.1   SIMULATION

We'll start by creating a relatively simple reverberant audio signal by passing the "raw" audio signal through an IIR having a substantial number of samples of delay (such that the equivalent time delay is about 50 msec or more), and a single feedback coefficient having a magnitude less than 1.0. In the z-domain such an IIR would have the transfer function

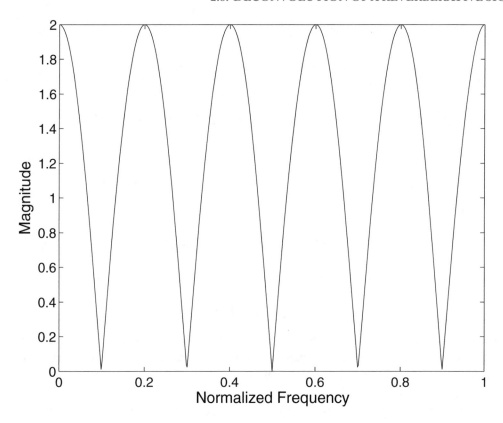

Figure 2.19: The frequency response of the net channel impulse response obtained as the convolution of the Channel Impulse Response [1,0,1] and a set of converged LMS tap weights which form a truncated inverse, [0,1,0,-1,0,1,0,-1,0,1].

$$H(z) = \frac{1}{1 - az^{-n}} \tag{2.2}$$

where a is a real number having magnitude less than 1 and n is chosen such that

$$Desired Echo Period = nT_s$$

where T_s is the sampling period.

Next, we'll estimate the values of a and n in Eq. (2.2) using only the reverberant signal itself, and then we'll design a filter which will undo (deconvolve) the original filter's work and return the audio signal to one without reverberation. The z-transform of such a filter is just the reciprocal of the transfer function that created the reverberant signal, namely

$$1 - az^{-n} \tag{2.3}$$

which is a simple FIR–a modified comb filter (i.e., one having $-1 < a < 1$).

The scripts (see exercises below)

$$LVxReverbDereverb(Delay, DecayRate, Mu, PeakSep)$$

and

$$LVxRevDerevShort(Delay, DecayRate, Mu, PeakSep, PrPC)$$

typical calls for which might be

$$\textbf{LVxReverbDereverb(1450,0.7,0.05,400)} \tag{2.4}$$

and

$$\textbf{LVxRevDerevShort(325,0.7,0.04,200,30)}$$

generate a reverberative audio signal using an IIR and then perform an analysis of the signal using autocorrelation to estimate the delay and adaptive filtering to estimate the decay rate. When these two parameters have been determined, the algorithm deconvolves the reverberative audio signal, returning the original audio signal without reverberation.

The parameter *Delay* corresponds to n in Eq. (2.2); *DecayRate* is the factor or coefficient a in Eq. (2.2), *Mu* is used in an LMS adaptive filtering process which estimates the value of *DecayRate*; and finally, *PeakSep* specifies the number of samples that must lie between any two adjacent peak values which are picked from the autocorrelation sequence to estimate the value of *Delay*. All of these variables will be discussed below in further detail.

Figure 2.20 shows the reverberative audio signal at (a), which was formed by convolving an audio signal (*drwatsonSR4K.wav* in this case) with an IIR having a single nonzero delay coefficient having *DecayRate* = 0.7 and *Delay* =1450 samples. At (b), the autocorrelation sequence of the reverberated audio signal is shown, and at (c), the dereverberated (or deconvolved) signal is shown. The peaks in the autocorrelation sequence are 1450 samples apart.

2.8.2   SPECTRAL EFFECT OF REVERBERATION

Now that we have both the original audio signal and the reverberated version, let's take a look at the spectral effect of imparting reverberation to the audio signal. One or more delayed versions of an audio signal, when mixed with the original signal at a given listening point (by acoustic superposition, for example) result in a distortion or filtering of the frequency spectrum of the composite audio signal. The same is true here, where the original audio signal is repeatedly mixed with itself at a basic delay of 1450 samples, for example.

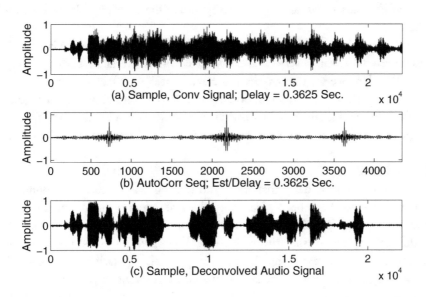

Figure 2.20: (a) Reverberative audio signal, formed by passing original audio signal through a simple IIR; (b) Autocorrelation sequence of reverberative audio signal; (c) Audio signal after being deconvolved using estimates of the IIR's delay and decay parameters using autocorrelation and LMS adaptive filtering.

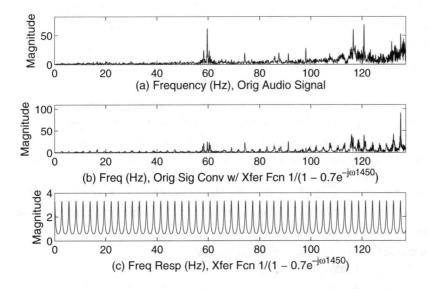

Figure 2.21: (a) Frequency response of original (nonreverberated) audio signal; (b) Frequency response of reverberated audio signal; (c) Frequency response (evaluated using the z-transform) of IIR filter used to generate the reverberative signal.

Plots (a) and (b) of Fig. 2.21 show the frequency response of the signal and the reverberated signal up to about 135 Hz.

It can be seen from Fig. 2.21, plots (b) and (c), that the IIR's frequency response is imparted to the audio signal, as expected. We can observe a comb filter-like effect. Recall that a comb filter is implemented with an FIR. Here, we see that an IIR can create a peaking effect with its poles, but the transfer function, without zeros in the numerator, does not drop to zero at any frequency. You may also notice that the IIR peaks are narrow, with wide valleys, whereas a comb filter effect has broad peaks and narrow valleys. One looks more or less like a vertically flipped version of the other.

2.8.3   ESTIMATING DELAY

Next, we'll discuss estimating the time delay that went into making the reverberative sound. We are pretending, of course, that we have been given only the reverberative audio signal and the information that it was created using a simple IIR. In a realistic situation, of course, the impulse response or process that creates the reverberative signal would be much more complex; it is only artificial reverberators that can create reverb signals that have perfectly regularly spaced echos. An actual acoustic environment generates, in general, an infinite number of irregularly spaced echos the delay times and attenuation factors of which depend on the room dimensions, the room contents, the locations of the source and listening point, and the reflective properties of the room and everything in it. It is very instructive, however, to see how one can estimate a basic delay and attenuation applying simple techniques that we have discussed previously.

The autocorrelation sequence is used extensively in speech analysis for generating an accurate frequency (pitch) estimate of a speech signal with a reasonable amount of computation. First, we compute the autocorrelation sequence of the reverberative signal, and then take the absolute value, as shown by the following m-code:

b = xcorr(ReverbSnd);

bAbs = abs(b);

Plot (b) of Fig. 2.20 shows a portion of the autocorrelation sequence as computed based on the m-code above. Figure 2.22, plot (a), shows just the most useful portion of the right half of that sequence. The distance in samples between major peaks is 1450. If the value of *DecayRate* is negative, every other major peak has a negative value, as shown in Fig. 2.22, plot (b), in which the value of *DecayRate* was -0.7 rather than +0.7.

These observations suggest a method of estimating the value of *Delay*, namely, computing the separation in samples between the two highest peaks in the absolute value of the autocorrelation sequence, the computation of which is performed by the following lines of m-code:

[locabsPeaks, absvalPeaks] = LVfindPeaks(bAbs,2,PeakSep);

absestDelay = abs(locabsPeaks(1) - locabsPeaks(2));

The first line of m-code above finds the two highest magnitude peaks and their locations in the signal *bAbs* which are at least *PeakSep* samples apart; in this case, *PeakSep* = 400. *PeakSep* is a parameter which can be estimated by eye if one doesn't have any idea of the actual value of *Delay*,

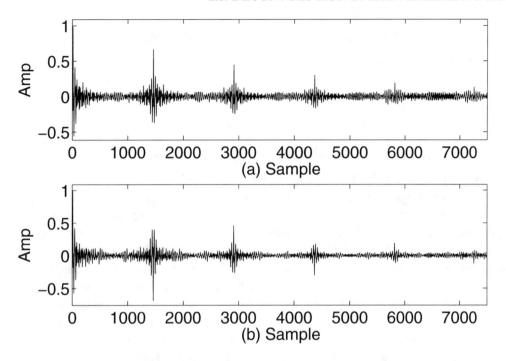

Figure 2.22: (a) Autocorrelation sequence of reverberated audio signal when DecayRate = +0.7; (b) Autocorrelation sequence of reverberated audio signal when DecayRate = -0.7.

which is what we are generally supposing is the case. For purposes of keeping things simple, though, the script allows you to specify the value of *PeakSep*, which you may do by knowing what value of *Delay* you are using. *PeakSep* must be less than the value of *Delay*. To estimate *PeakSep* by eye, you would plot the autocorrelation sequence and examine it for peaks, noting the distance between what appear to be suitable candidate peaks, and then perform the estimation of *Delay* by calling *LVfindPeaks* with *PeakSep* specified as about half the observed candidate peak separation (details of the function *LVfindPeaks* are in the exercises below).

2.8.4   ESTIMATING DECAYRATE

The process that created *ReverbSnd* was an IIR filter; therefore, the inverse or deconvolution process is an FIR filter the z-transform of which is the reciprocal the z-transform of the IIR, as shown above. We have obtained the value for *Delay*, so the z-transform would be

$$\frac{Output(z)}{Input(z)} = 1 - az^{-Delay}$$

which leads to

$$Output(z) = Input(z) - a \cdot Input(z) \cdot z^{-Delay}$$

and after conversion to a difference equation yields

$$Output[n] = Input[n] - a \cdot Input[n - Delay] \qquad (2.5)$$

or in m-code

DeconOutput(Ctr) = ReverbSnd(Ctr)-a*ReverbSnd(Ctr-estDelay)

This can be treated as an adaptive process in which the deconvolved output, $DeconOutput(Ctr)$, is considered as an error signal arising from the difference between a signal $ReverbSnd(Ctr)$ which is the convolution of the original signal and an unknown system's impulse response, and an adaptive filter output $a \cdot ReverbSnd(Ctr - estDelay)$ in which there is a single weight a to be determined. We are taking an earlier sample from $ReverbSnd$ (earlier by $estDelay$) and trying to find the proper weighting coefficient a which will minimize the error. This works well when the later sample of $ReverbSnd$ has no signal content except the delayed or reverberated earlier sound, which we are attempting to remove. When the later sample already has content (which is usually the case, except for pauses between words), the situation is much the same as in echo cancelling with Near End speech: there is a signal contributing to the error signal, but which is not also present in the filter input. Thus, a Dual-H system is warranted, which automatically detects the absence of content (equivalent to no Near End speech in echo cancelling) by yielding a high value of the ratio of power into the adaptive filter (in this case samples of the reverberated sound) to power output from the error junction. This is similar to the ERLE and IRLE parameters discussed earlier in the chapter; here, we'll call it *Reverb Return Loss Enhancement*, or *RRLE*.

The method we are employing to remove the echo or reverberation from the original signal is very analogous to sparse echo cancellation, which was discussed earlier in the chapter. Note that we used correlation to estimate the relevant tap delay, which was (based on a certain amount of knowledge of the echo or reverberative process), a single delay. We then use only the one tap in the adaptive filter. If we had estimated the delay at, say, 900 samples, we could have used an adaptive filter having, say, 1000 active taps, covering any echo components occurring in the time delay range of 0 to 1000 samples. This, however, not only would result in far more computational overhead, but would likely produce inferior results.

Figure 2.23, plot (a), shows the sequence of estimates of the single coefficient being estimated, the value of *DecayRate* during a trial run of script call (2.4), while plot (b) shows the corresponding sequence of values of *RRLE*.

2.8.5   DECONVOLUTION

After both *Delay* and *DecayRate* have been estimated, the values are used in a deconvolution filter which has as its z-transform the reciprocal of the z-transform of the IIR which was used to generate the reverberated audio signal. The relevant lines of m-code are:

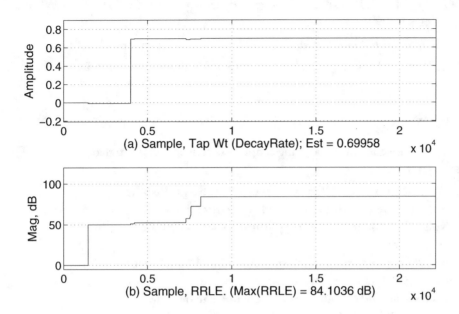

Figure 2.23: (a) Estimated value of DecayRate (the amplitude of the pole of the IIR used to form the reverberative audio signal); (b) Reverb Return Loss Enhancement (RRLE), used to tell the LMS Adaptive filter when to update the coefficient.

> **deconImp = [1,zeros(1,estDelay-1),-estDecayRate];**
> **DereverbSnd = filter(deconImp,1,ReverbSnd)**

If all has gone well, the dereverberated audio signal will be identical or nearly so to the original audio signal.

2.9   REFERENCES

[1] Bernard Widrow and Samuel D. Stearns, *Adaptive Signal Processing*, Prentice-Hall, Englewood Cliffs, New Jersey, 1985.

[2] Paul M. Embree and Damon Danieli, *C++ Algorithms for Digital Signal Processing*, Prentice-Hall PTR, Upper Saddle River, New Jersey, 1999.

[3] Ali H. Sayed, *Fundamentals of Adaptive Filtering*, John Wiley & Sons, Hoboken, New Jersey, 2003.

[4] John R. Buck, Michael M. Daniel, and Andrew C. Singer, *Computer Explorations in Signals and Systems Using MATLAB, Second Edition*, Prentice-Hall, Upper Saddle River, New Jersey 07548, 2002.

[5] Vinay K. Ingle and John G. Proakis, *Digital Signal Processing Using MATLAB V.4*, PWS Publishing Company, Boston, 1997.

[6] U. S. Patent 7,107,303 to Kablotsky et al, *Sparse Echo Canceller*, September 12, 2006.

[7] U. S. Patent 7,181,001 to Demirekler et al, *Apparatus and Method for Echo Cancellation*, February 20, 2007.

2.10   EXERCISES

1. Write the m-code to implement the following script, as described and illustrated in the text. Test it using at least the given test calls.

> **function LVxLMSANCNorm(PlantCoeffVec,k,Mu,freq,DVMult)**
> **% This script implements a 10-tap LMS adaptive filter in an**
> **% Active Noise Cancellation (ANC) application, such as cancell-**
> **% ing noise in a duct.**
> **% PlantCoeffVec is a row vector of 10 coefficients for the FIR that**
> **% simulates the Plant (i.e., the Duct impulse response).**
> **% k specifies the standard deviation of random noise to be mixed**
> **% with two sinusoids having frequencies freq and 3*freq and**
> **% respective amplitudes of 1.0 and 0.5;**
> **% Mu specifies the tap weight update damping coefficient;**
> **% An NLMS algorithm is used; its effectiveness can be tested**
> **% by giving the test signal various stepped-amplitude profiles**
> % with the input argument DVMult, which is a vector of
> % amplitudes to impose on the test signal as a succession of
> % equally spaced amplitude segments over a test signal of 15
> % times the filter length.
> **% Test calls:**
> **% LVxLMSANCNorm([0,0,1,0,-0.5,0.6,0,0,-1.2,0],2,2,27,[1,2,5,8])**
> **% LVxLMSANCNorm([0,0,1,0,-0.5,0.6,0,0,-1.2,0],2,0.2,27,[1,2,5,8])**
> **% LVxLMSANCNorm([0,0,1,0,-0.5,0.6,0,0,-1.2,0],0,2,3,[1,2,5,8])**

2. Write the m-code to implement the following script, as described and illustrated in the text:

> **function LVxModelPlant(A, LenLMS, NoPrZs, NoPrPs, Mu,...**
> **tSig, NAmp, NoIts)**
> **% A and B are a set of IIR and FIR coefficients to generate the**
> **% Plant response LenLMS is the number of LMS adaptive FIR**
> **% coefficients to use to model the IIR; it should generally be long**
> **% enough to model the impulse response until it has mostly**
> **% decayed away**
> **% NoPrZs is the number of Prony zeros to use to model the**

% LMS-derived impulse response (the converged LMS adaptive
% filter coefficients, which form a truncated version of the Plant
% impulse response); NoPrPs is the number of Prony poles to use
% to model the LMS-derived impulse response; Mu is the standard
% LMS update term weight; tSig if passed as 0 = white noise;
% 1 = DC; 2 = Nyquist; 3 = Half-band; 4 = mixed cosines (DC,
% Half-band, and Nyquist)
% NAmp = amplitude of noise to add to selected tSig
% NoIts is the number of iterations to do in the adaptive process.
% NoIts should be at least 10 times LenLMS.

Test the script with at least the following calls:

% LVxModelPlant([1,-0.9],[1],100, 2,2, 0.5,0,0,1000)
% LVxModelPlant([1, 0,0.81],[1],100, 3,3,0.5,1, 0,1000)
% LVxModelPlant([1,-1.3,0.81],[1],100, 3,3,0.5,1, 0,1000)
% LVxModelPlant([1,-1,0.64],[1],100, 3,3, 0.5, 1, 0,1000)

3. Write the m-code for the following script, which provides a basic illustration of echo cancellation using Single- and Dual-H methods, using a user-selectable percentage of the reduced-bandwidth audio files 'drwatsonSR4K.wav' and 'whoknowsSR4K.wav'. Detailed instructions are given after the function definition.

```
function LVxLMSAdptFiltEchoShort(k,Mu,DHorSH,...
MuteNrEnd,PrPC)
% k is the amount of noise to add to the Far End signal (see below)
% Mu is the usual LMS update term weight;
% DHorSH yields a Dual-H system if passed as 1, or a Single-H
% system if passed as 0; MuteNrEnd mutes the desired signal
% ('drwatsonSR4K') at its very beginning if passed as 1, about
% one-third of the way through if passed as 2, or no muting at
% all if passed as 0.
% The Near End signal consists of the audio file 'drwatsonSR4K.wav'
% The Far End signal consists of the audio file 'whoknowsSR4K.wav'
% plus random noise weighted by k, limited to the length of the
% Near End signal;
% The total amount of computation can be limited by processing
% only a percentage of the audio signals, and that percentage
% is passed as the input variable PrPC.
% Using whoknowsSR4K.wav and drwatsonSR4K.wav as test signals,
% 100%=22184 samps. The adaptive FIR is ten samples long
% and the echo is simulated as a single delay of 6 samples,
% for example. The final filtered output signal is interpolated
```

% by a factor of 2 using the function interp to raise its sample
% rate to 8000 Hz so it may be played using the call sound
% (EchoErr,8000) after making the call global EchoErr in
% the Command window. A single figure is created showing
% the filtered output signal
% Test calls:
% LVxLMSAdptFiltEchoShort(0.1,0,0,0,50) % Mu = 0, FarEnd hrd
% LVxLMSAdptFiltEchoShort(0.1,0.2,0,0,50) % S-H, no Mute
% LVxLMSAdptFiltEchoShort(0.1,0.2,0,1,50) % S-H, Mute Imm
% LVxLMSAdptFiltEchoShort(0.1,0.2,1,0,50) % D-H, no Mute
% LVxLMSAdptFiltEchoShort(0.1,0.2,1,1,50) % D-H, Mute Imm
% LVxLMSAdptFiltEchoShort(0.1,0.2,1,2,50) % D-H, Mute del
% LVxLMSAdptFiltEchoShort(0.01,0.2,0,0,50) % S-H, no Mute
% LVxLMSAdptFiltEchoShort(0.01,0.2,0,1,50) % S-H, Mute Imm
% LVxLMSAdptFiltEchoShort(0.01,0.2,1,0,50) % D-H, no Mute
% LVxLMSAdptFiltEchoShort(0.01,0.2,1,1,50) % D-H, Mute Imm
% LVxLMSAdptFiltEchoShort(0.01,0.2,1,2,50) % D-H, Mute del
% LVxLMSAdptFiltEchoShort(0.01,0.005,1,0,60) % Low Mu, can
% hear Far End gradually diminish

Since this script is intended to execute as quickly as possible, the detection of Near End speech has been eliminated, and archiving and plotting of ERLE as well. The code should be written to automatically play the resultant filtered output at least once. Insert in the m-code the statement

global EchoErr

where $EchoErr$ is the filtered output signal. Then, in the Command Window, type **global EchoErr** after the program has run, and you can then play the filtered output at will by making the call

sound(EchoErr, 8000)

In order to avoid retaining older (unwanted) computed values in a global variable, it is good practice is to insert the global statement early in the script prior to computing and assigning new values to the variable, and to follow the global statement immediately with a statement to empty the variable, i.e.,

global EchoErr
EchoErr = [];

Since the net filtered output is the delayed Far End Signal plus the Near End signal, minus the FIR output, the delayed Far End signal can be added to the Near End Signal prior to the loop filtering operation, resulting in only one array access rather than two to compute the filtered output. The coefficient update normalizing term should be two times Mu, divided by the sum of the squares of the signal in the FIR, and this, too, can be computed ahead of the loop and looked up while in the filtering loop.

Since the audio files were sampled at 4 kHz, the final filtered output signal must be interpolated by a factor of two so that its net final sample rate is 8000 Hz (LabVIEW only allows certain sample rates to be used in the *sound* function, namely, 8000, 11025, 22050, and 44100 Hz).

4. Write the m-code to implement the script *LVxLMSAdaptFiltEcho*, as described and illustrated in the text. The preceding exercise created a much simpler script, which does not use the variables *DblTkUpD*, *SampsVOX*, *VOXThresh*, and which requires a much smaller computation load. This project adds these variables, and archives ERLE and several other variables for purposes of plotting. The preceding project can serve as a good starting basis for this project.

```
function LVxLMSAdaptFiltEcho(k,Mu,DHorSH,MuteNrEnd,...
DblTkUpD,SampsVOX,VOXThresh)
% k is the amount of noise to add to the interference signal
% ('drwatsonSR8K.wav'); Mu is the usual LMS update term
% weight; DHorSH yields a Dual-H system if passed as 1, or
% a Single-H system if passed as 0; MuteNrEnd mutes the
% desired signal ('drwatsonSR8K.wav') at its very beginning
% if passed as 1, about one-third of the way through if passed
% as 2, or no muting at all if passed as 0. In Single-H mode,
% DblTkUpD, if passed as 0, allows coefficient updating any
% time, but if passed as 1 prevents coefficient update when
% the most recent SampsVOX samples of the desired (Near
% End) signal level are above VOXThresh. In Dual-H mode,
% DblTkUpD,SampsVOX,VOXThresh are ignored.
% The Far End signal is the audio file 'whoknowsSR8K.wav',
% limited to the length of 'drwatsonSR8K/wav'
% The adaptive FIR is ten samples long and the echo is
% simulated as a single delay of 6 samples, for example.
% A figure having four subplots is created. The first subplot
% shows the values of all ten FIR coefficients over the entire
% course of computation; the second subplot shows the filtered
% output signal; the third subplot shows as a binary plot
% whether or not coefficient update was allowed at any given
% stage of the computation based on the detection of Near
% End speech; the fourth subplot shows the figure of merit
% (ERLE, as defined in the text) over the course of the
% computation. The filtered output audio signal may be played
% using the call sound(EchoErr,8000) after making the call
% global EchoErr in the Command window. The Near End plus
```

% **echoed Far End signal (the signal entering the error junction)**
% **can be played with the call sound(NrEndPlusFarEnd,8000)**
% **after making the call global NrEndPlusFarEnd.**

Test the script with the following calls, then repeat the calls, changing k to 0.2; compare the audibility of the Far End signal between the same calls using different amounts of noise as specified by k.

% **LVxLMSAdaptFiltEcho(0.02,0.2,1,0,[],[],[]) % D-H, no Mute**
% **LVxLMSAdaptFiltEcho(0.02,0.2,1,1,[],[],[]) % D-H, Mute Im**
% **LVxLMSAdaptFiltEcho(0.02,0.2,1,2,[],[],[]) % D-H, del Mute**
% **LVxLMSAdaptFiltEcho(0.02,0.2,0,0,0,50,0.05) % S-H, dbltalk**
% **update, no Mute**
% **LVxLMSAdaptFiltEcho(0.02,0.2,0,1,1,50,0.05) % S-H, no**
% **dbltalk update, Mute Im**

5. Write the m-code to implement the following script, as described and illustrated in the text:

function LVxLMSAdaptFiltDeCorr(k,Mu,freq,NoTaps,...
DeCorrDelay)
% **k is an amount of white noise to add to a test**
% **signal which consists of a sinusoid of frequency freq;**
% **Mu is the LMS update term weight;**
% **NoTaps is the number of taps to use in the LMS adaptive filter;**
% **DeCorrDelay is the number of samples to delay the filter input**
% **relative to the Plant or channel delay.**

Test the script with at least the following calls:

% **LVxLMSAdaptFiltDeCorr(0.1,0.75,200,20,5)**
% **LVxLMSAdaptFiltDeCorr(0.1,0.75,200,20,0)**

6. Write the m-code to implement the script *LVxLMSInterferCancShort* as specified below.

function LVxLMSInterferCancShort(k,Mu,DHorSH,MuteDesSig,...
GP2,GP4,DP1,DP2,DP3,DP4,NumTaps,PrPC)
% **k is the amount of noise to add to the interference signal**
% **('whoknowsSR4K.wav'); Mu is the usual LMS**
% **update term weight;**
% **DHorSH yields a Dual-H system if passed as 1, or a Single-H**
% **system if passed as 0. MuteDesSig mutes the desired signal**
% **('drwatsonSR4K.wav') at its very beginning if passed as 1,**
% **about one-third of the way through if passed as 2, or no muting**
% **at all if passed as 0; GP2 is the gain in the path from the desired**
% **sound source to the noise reference microphone, and GP4 is the**
% **gain in the path from the noise source to the desired signal**

% microphone;
% DP1 is the Delay in samples from the desired sound source to
% the desired sound source microphone; DP2 is the Delay from
% the desired sound source to the noise microphone; DP3 is the
% Delay from the noise source to the noise source microphone;
% DP4 is the Delay from the noise source to the desired sound
% microphone. NumTaps is the number of Delays or taps to use
% in the adaptive filter. The total amount of computation can be
% limited by processing only a percentage of the audio signals,
% and that percentage is passed as the input variable PrPC.
% Using whoknowsSR4K.wav and drwatsonSR4K.wav as test
% signals, 100%=22184 samps. The final filtered output signal
% is interpolated by a factor of 2 using the function interp to
% raise its sample rate to 8000 Hz so it can be played using the
% call sound(ActualErr,8000) after running the script and making
% the call global ActualErr in the Command window
% Typical calls might be:
% LVxLMSInterferCancShort(0.02,0.3,1,0,1,1,1,6,1,6,5,30)
% LVxLMSInterferCancShort(0.02,0.3,1,0,0.16,0.16,...
7,42,7,42,35,30)
% LVxLMSInterferCancShort(0.02,0.3,1,0,0.06,0.06,...
1,6,1,6,5,30)
% LVxLMSInterferCancShort(0.02,0.3,1,0,1,1,6,1,1,6,5,30)
% LVxLMSInterferCancShort(0.02,0.3,1,1,1,1,1,6,1,6,5,30)
% LVxLMSInterferCancShort(0.02,0.2,0,2,1,1,1,6,1,6,5,30)
% LVxLMSInterferCancShort(0.02,0.3,0,1,1,1,1,6,1,6,5,30)

7. Write the m-code to implement the following script, as described and illustrated in the text:

function LVxLMSInterferCancel(k,Mu,DHorSH,MuteDesSig,...
DblTkUpD,NoSampsVOX,...
% VOXThresh,GP2,GP4,DP1,DP2,DP3,DP4,NumTaps)
% k is the amount of noise to add to the interference signal
% ('whoknowsSR8K.wav'). Mu is the usual LMS update term
% weight; DHorSH yields a Dual-H system if passed as 1, or
% a Single-H system if passed as 0. MuteDesSig mutes the
% desired signal ('drwatsonSR8K.wav') at its very beginning
% if passed as 1, about one-third of the way through if passed
% as 2, or no muting at all if passed as 0; DblTkUpD, if passed
% as 0, allows coefficient updating any time, but if passed as 1
% prevents coefficient update when the most recent NoSamps

% VOX samples of the desired signal level are above VOXThresh.
% GP2 is the gain in the path from the desired sound source to
% the noise reference microphone, and GP4 is the gain in the
% path from the noise source to the desired signal microphone;
% DP1 is the Delay in samples from the desired sound source
% to the desired sound source microphone;
% DP2 is the Delay from the desired sound source to the noise
% microphone; DP3 is the Delay from the noise source to the
% noise source microphone; DP4 is the Delay from the noise
% source to the desired sound microphone. NumTaps is the
% number of Delays or taps to use in the adaptive filter.
% The filtered audio output can be played using the call
% sound(ActualErr,8000) after running the script and making
% the call global ActualErr in the Command window.

Test the script with at least the following calls:

% LVxLMSInterferCancel(0.02,0.3,1,0,0,50,0.03,1,1,1,6,1,6,5)
% LVxLMSInterferCancel(0.02,0.3,1,0,0,50,0.03,0.16,...
0.16,7,42,7,42,35)
% LVxLMSInterferCancel(0.02,0.3,1,0,1,50,0.03,0.06,...
0.06,1,6,1,6,5)
% LVxLMSInterferCancel(0.02,0.3,1,0,0,50,0.03,1,1,6,1,1,6,5)
% LVxLMSInterferCancel(0.02,0.3,1,1,0,50,0.03,1,1,1,6,1,6,5)
% LVxLMSInterferCancel(0.02,0.2,0,2,1,80,0.03,1,1,1,6,1,6,5)
% LVxLMSInterferCancel(0.02,0.3,0,1,1,50,0.03,1,1,1,6,1,6,5)

8. A certain signal transmission channel has the transfer function

$$H(z) = 1 + 0.9z^{-1} + 0.8z^{-2}$$

Under ideal convergence conditions, what would be the shortest adaptive FIR that could, when well-converged, result in an equalized channel transfer function having a magnitude response within 0.1 dB of unity?

9. Write the m-code to implement the following script, as described and illustrated in the text, creating a figure with subplots as shown, for example, in Fig. 2.17:

function LVxChannEqualiz(ChImp,Mu,tSig,lenLMS)
% ChImp is the impulse response of the channel which is to
% be equalized by the LMS FIR;
% Mu is the usual update weight
% tSig designates the type of test signal to be used:
% 0 = random noise
% 1 = Unit impulse sequence of length lenLMS, repeated at
% least ten times, with alternating sign, +, -, +, etc
% 2 = Unit impulse sequence of length lenLMS, repeated at
% least ten times
% 3 = Sinusoid at the half-band frequency
% lenLMS is the length of the adaptive LMS FIR

Try two different values for the delay (as shown in Fig. 2.16), namely, (1) one-half the sum of the lengths of the Channel Impulse Response and the LMS FIR, and (2) one-half the length of the Channel Impulse Response. In both cases, round the value obtained in (1) or (2) to the nearest integral value. Test the script with at least the following calls, trying each of the two delay values to obtain the better response.

% LVxChannEqualiz([-0.05,0,-0.5,0,-1,0,1,0,0.5,0,0.05],1.5,2,131)
% LVxChannEqualiz([1,0,1],2.1,2,91) % notch
% LVxChannEqualiz([1,0,1],2.4,2,191) % notch
% LVxChannEqualiz([1 1],2.2,2,65) % lowpass
% LVxChannEqualiz([1,0,-1],1.8,2,85) % bandpass
% LVxChannEqualiz([1,-1],2.5,2,179) % highpass
% LVxChannEqualiz([1,-0.8],2.3,2,35) % highpass
% LVxChannEqualiz([1,0.8],2.4,2,35) % lowpass
% LVxChannEqualiz([0.3,1,0.3],2.3,2,9) % lowpass
% LVxChannEqualiz([1,0.7],2.3,2,25) % lowpass
% LVxChannEqualiz([1,0,0,0,0,0,1],1.6,2,179) % comb filter
% LVxChannEqualiz([1,0,0,-0.3,0,0,0.05],2.5,2,23)
% LVxChannEqualiz([1,0,-1],1.5,2,55)
% LVxChannEqualiz(([1,0,1,0,1,0,1].*hamming(7)'),2.1,2,91)

10. Write the m-code for the script *LVxRevDerevShort*, which is a shorter version of the script *LVxReverbDereverb* (see exercise immediately following).

function LVxRevDerevShort(Delay,DecayRate,Mu,PeakSep,PrPC)
% Delay is the number of samples of delay used in the single
% feedback stage to generate the reverberated sound file.
% DecayRate is a real number which is feedback gain.
% Mu is the usual LMS update weight term.
% PeakSep the number of samples by which to separate

% detected peaks in the autocorrelation sequence of the
% reverberated sound. Uses a Dual-H architecture to estimate
% DecayRate, and uses the best estimate of DecayRate to
% block filter the reverberated sound to produce the
% dereverberated sound. Does not archive or plot the values
% of RRLE and current best estimate or DecayRate. Plays
% the reverberated sound once and the dereverberated
% sound once. These sounds may be played again after
% running the script by making the calls global ReverbSnd
% and global DereverbSnd in the Command window and then
% making the calls sound(ReverbSnd,8000) and sound
% (DereverbSnd,8000) at will. Processes only a percentage
% of the audio file equal to PrPC in the adaptive filtering
% loop to estimate DecayRate. Uses the audio file
% 'drwatsonSR4K.wav' as the test signal and at the end
% both the reverberative sound and the dereverberated
% sound are interpolated to a sample rate of 8000 Hz to
% be played by the function sound(x,Fs),which only allows
% Fs = 8000, 11025, 22050, or 44100.
% Sample call:
% LVxRevDerevShort(325,0.7,0.05,200,30)
% LVxRevDerevShort(225,0.7,0.12,200,30)
% LVxRevDerevShort(625,0.7,0.06,200,47)
% LVxRevDerevShort(1625,0.7,0.07,200,47)

You will need to extract peak locations for the autocorrelation; suitable code is found below, and the script is provided with the basic software for this book. Figure 2.24 shows the result of the call

$$[Locs,Pks] = LVfindPeaks((cos(2*pi*2*[0:1:63]/64)),3,16)$$

with the highest three peaks marked with circles. Note that a zone of ineligible values is set up around each peak that has already been chosen, otherwise peaks might be chosen right next to each other when that is not the desire.

One thing that makes the adaptive LMS algorithm run much more quickly is to recognize the fact that the times when RRLE should increase are when the Near End signal is close to zero in magnitude; such times form a contiguous set of samples. Thus, in stepping through the algorithm, it can be done by using every fifth sample, for example, rather than every sample, and whenever RRLE increases, the algorithm goes back to computing the update coefficients for every sample rather than every fifth sample. In this way, only a little more than 20% of the samples are used to compute the next estimate of *DecayRate* and the corresponding value of RRLE.

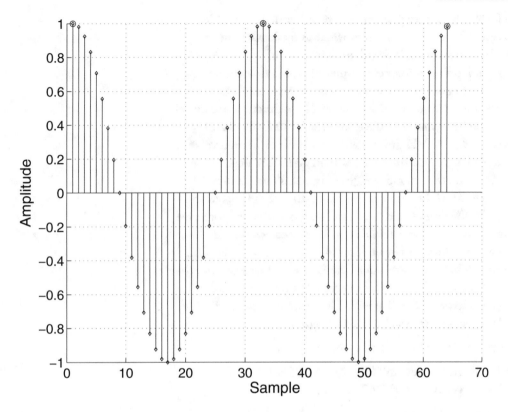

Figure 2.24: Two cycles of a cosine wave over 64 samples with the highest 3 peaks marked with circles (see text).

function [Locs, Pks] = LVfindPeaks(Mat,QuanPks,PeakSep)
% Mat is a vector or matrix for which QuanPks peak values
% **are sought, each of which is at least PeakSep samples**
% **distant from the next closest peak value. Locs tells the**
% **locations of the peak values in the matrix. To make adjacent**
% **samples of Mat eligible to be peaks, use PeakSep = 0.**
% **Sample call:**
% **[Locs, Pks] = LVfindPeaks([0:1:7],8,0)**
% **[Locs, Pks] = LVfindPeaks([0:1:7],8,1)**
% **[Locs, Pks] = LVfindPeaks((cos(2*pi*3*[0:1:63]/64)),4,8)**

11. Write the m-code to implement the following script, as described and illustrated in the text:

function LVxReverbDereverb(Delay,DecayRate,Mu,PeakSep)
% Delay is the number of samples of delay used in the single

% **feedback stage to generate the reverberated sound file.**
% **DecayRate is a real number which is feedback gain.**
% **Mu is the usual LMS update weight term.**
% **PeakSep the number of samples by which to separate**
% **detected peaks in the autocorrelation sequence of the**
% **reverberated sound. Uses a Dual-H architecture to estimate**
% **DecayRate, and uses the best estimate of DecayRate**
% **from the Dual-H process along with the estimated value**
% **of Delay (from autocorrelation) in an FIR to deconvolve**
% **the entire reverberative signal. Archives and plots the**
% **values of RRLE and current best estimate of DecayRate.**
% **Plays the dereverberated sound once and the reverberative**
% **sound once. These sounds may be played again after**
% **running the script by making the calls global ReverbSnd**
% **and global DereverbSnd in the Command window and then**
% **making the calls sound(ReverbSnd,8000) and sound**
% **(DereverbSnd,8000) at will. The script uses the audio file**
% **'drwatsonSR8K.wav' as the test signal.**

Test the script with at least the following calls:

% **LVxReverbDereverb(650,0.7,0.05,400)**
% **LVxReverbDereverb(3600,0.975,0.05,400)**

12. Write a script that generates a test signal by reading the audio file 'drwatsonSR8K.wav' and adding an interfering sinusoid having amplitude *A* and frequency *Freq* to it. The script will then use LMS adaptive filtering with a decorrelating delay to remove the sinusoidal tone from the audio signal. Code suitable to create the test signal is

[y,Fs,bits] = wavread('drwatsonSR8K.wav');
y = y'; lenFile = length(y);
y = (1/max(abs(y)))*y;
t = [0:1:lenFile-1]/Fs;
LMSDeCorDataVec = y + A*sin(2*pi*t*Freq);

The display figure for the script should be like Fig. 2.25, including four subplots that show the input signal (audio file corrupted by a large amplitude sinusoid), the filtered output signal, and the magnitude of the spectrum of each, computed by FFT. Figure 2.25 was generated by making the call

LVxLMSDeCorrHeterodyne('drwatsonSR8K.wav',1.5,0.08,500,45,1)

Your script should conform to the syntax below; test the script with the given test calls.

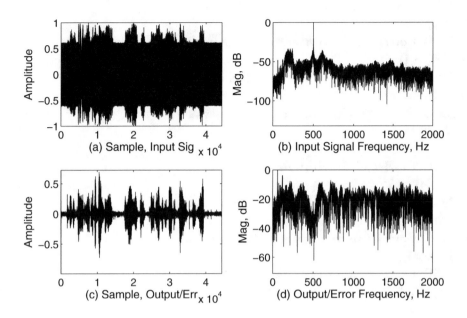

Figure 2.25: (a) Input signal, audio file 'drwatsonSR8K' corrupted with a large amplitude sinusoid; (b) Spectrum of signal in (a); (c) Filtered (output or error) signal, showing original audio signal with no visible interfering sinusoid; (d) Spectrum of signal in (c). Note the deep spectral notch at the frequency of the interfering sinusoid as shown in plot (b).

```
function LVxLMSDeCorrHeterodyne(strWavFile,A,Mu,Freq,...
NoTaps,DeCorrDelay)
% strWavFile is the audio file to open and use as the test signal
% A is the amplitude of an interfering tone (sinusoid) to be added
% to the audio file whose file name or path name (including file
% name and extension) is specified as strWavFile. Mu is the LMS
% update term weight; NoTaps is the number of taps to use in
% the LMS adaptive filter; DeCorrDelay is the number of samples
% to delay the filter input relative to the Plant or channel delay.
% The following plots are created: Frequency response (DTFT)
% of the adaptive filter when converged (magnitude v. frequency
% in Hz), the test signal (amplitude v. sample), the filtered test
% signal (amplitude v. sample), the DTFT of the test signal
% (magnitude v. frequency in Hz, and the DTFT (magnitude v.
% frequency in Hz) of the last 50 % of the filtered test signal..
% Test calls:
```

```
% LVxLMSDeCorrHeterodyne('drwatsonSR8K.wav',...
0.5,0.08,250,45,1)
% LVxLMSDeCorrHeterodyne('drwatsonSR8K.wav',...
0.5,0.08,500,45,1)
% LVxLMSDeCorrHeterodyne('drwatsonSR8K.wav',...
0.5,0.08,1200,45,1)
% LVxLMSDeCorrHeterodyne('drwatsonSR8K.wav',...
0.5,0.01,75,45,10)
%
% LVxLMSDeCorrHeterodyne('drwatsonSR8K.wav',...
1.5,0.08,250,45,1)
% LVxLMSDeCorrHeterodyne('drwatsonSR8K.wav',...
1.5,0.08,500,45,1)
% LVxLMSDeCorrHeterodyne('drwatsonSR8K.wav',...
1.5,0.08,1200,45,1)
% LVxLMSDeCorrHeterodyne('drwatsonSR8K.wav',...
1.5,0.01,75,45,10)
%
% LVxLMSDeCorrHeterodyne('drwatsonSR8K.wav',...
4.5,0.08,250,45,1)
% LVxLMSDeCorrHeterodyne('drwatsonSR8K.wav',...
4.5,0.08,500,45,1)
% LVxLMSDeCorrHeterodyne('drwatsonSR8K.wav',...
4.5,0.08,1200,45,1)
% LVxLMSDeCorrHeterodyne('drwatsonSR8K.wav',...
4.5,0.01,75,45,10)
%
% LVxLMSDeCorrHeterodyne('drwatsonSR8K.wav',...
0.02,0.08,250,45,1)
% LVxLMSDeCorrHeterodyne('drwatsonSR8K.wav',...
0.02,0.08,500,45,1)
% LVxLMSDeCorrHeterodyne('drwatsonSR8K.wav',...
0.02,0.08,1200,45,1)
% LVxLMSDeCorrHeterodyne('drwatsonSR8K.wav',...
0.02,0.01,75,45,10)
```

Use the script *LVx_AnalyzeModWavFile* (developed in the exercises for the chapter devoted to FIR filter design methods in Volume III of the series) to make the test file

$$drwat8Kplus400HzAnd740Hz.wav$$

for use in the following test call. The amplitudes of the 400 Hz and 740 Hz waves should be specified as 0.02 when creating the file *'drwat8Kplus400HzAnd740Hz.wav'* using the script *LVx_AnalyzeModWavFile*.

> % **LVxLMSDeCorrHeterodyne('drwat8Kplus400HzAnd740Hz.wav',...**
> **0,0.01,0,50,10)**
> % **LVxLMSDeCorrHeterodyne('drwat8Kplus400HzAnd740Hz.wav',...**
> **0.02,0.01,1100,50,10)**

13. Write a script that is essentially the same as *LVxLMSDeCorrHeterodyne* with the exception that the new script will use a Dual-H architecture. You may notice that the script *LVxLMSDeCorrHeterodyne* does not work well unless the persistent tone is of relatively high amplitude. When the amplitude of the persistent tone is relatively low, performance of the adaptive filter is poor, leading to distortion and inadequate removal of the persistent tone. You should find that the Dual-H architecture will function much better with low level persistent tones than does the simple Single-H architecture of the script *LVxLMSDeCorrHeterodyne*. The test calls presented with the function specification below are presented in pairs, with the same parameters being used in both scripts, *LVxLMSDeCorrHeterodyne* and *LVxLMSDeCorrHetDualH*, for comparison.

> **function LVxLMSDeCorrHetDualH(strWavFile,A,Mu,Freq,...**
> **NoTaps,DeCorrDelay)**
> % **Has the same input arguments and function as**
> % **LVxLMSDeCorrHeterodyne, but uses a Dual-H**
> % **architectures for better performance.**
> % **strWavFile is the audio file to open and use as the test signal**
> % **A is the amplitude of an interfering tone (sinusoid) to be**
> % **added to the audio file whose file name or path name**
> % **(including file name and extension) is specified as**
> % **strWavFile.**
> % **Mu is the LMS update term weight;**
> % **NoTaps is the number of taps to use in the LMS adaptive filter;**
> % **DeCorrDelay is the number of samples to delay the filter input**
> % **relative to the Plant or channel delay.**
> % **The following plots are created: Frequency response (DTFT)**
> % **of the adaptive filter when converged (magnitude v.**
> % **frequency in Hz), the test signal (amplitude v. sample),**
> % **the filtered test signal (amplitude v. sample), the DTFT**
> % **of the test signal (magnitude v. frequency in Hz, and**
> % **the DTFT (magnitude v. frequency in Hz) of the last**
> % **50 % of the filtered test signal.**
> %
> % **Test calls (pairs to be compared with each other):**

%

% **LVxLMSDeCorrHeterodyne('drwatsonSR8K.wav',...**
0.012,0.03,850,65,7)

% **LVxLMSDeCorrHetDualH('drwatsonSR8K.wav',...**
0.012,0.03,850,65,7)

%

% **LVxLMSDeCorrHeterodyne('drwat8Kplus400HzAnd740Hz.wav',...**
0,0.025,0,73,3)

% **LVxLMSDeCorrHetDualH('drwat8Kplus400HzAnd740Hz.wav',...**
0,0.025,0,73,3)

%

% **LVxLMSDeCorrHeterodyne('drwat8Kplus400HzAnd740Hz.wav',...**
0.02,0.025,1150,73,3)

% **LVxLMSDeCorrHetDualH('drwat8Kplus400HzAnd740Hz.wav',...**
0.02,0.025,1150,73,3)

The calls

LVxLMSDeCorrHeterodyne('drwat8Kplus400HzAnd740Hz.wav',...
0.04,0.025,1150,73,3)
LVxLMSDeCorrHetDualH('drwat8Kplus400HzAnd740Hz.wav',...
0.04,0.025,1150,73,3)

result in Figs. 2.26 and 2.27, respectively, which show the spectrum of the test signals and filtered test signals for the Single-H and Dual-H versions of the persistent-tone removal script.

The call

> **LVxLMSDeCorrHetDualH('drwat8Kplus400HzAnd740Hz.wav',...**
> **0.04,0.025,1150,400,7)**

results in Fig. 2.28, in addition to the other figures specified in the function description. Note the peaks in the filter response at the persistent frequencies, and recall that the filtered signal is derived by subtracting the LMS filter output from the test signal.

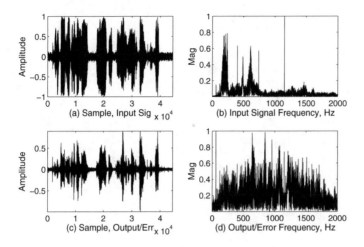

Figure 2.26: (a) Test speech signal having persistent tones added at frequencies of 440 Hz, 740 Hz, and 1150 Hz; (b) Spectrum of test signal in (a); (c) Output/Error signal, i.e., the test signal after being filtered by a Single-H LMS system having a decorrelating delay designed to assist in removing persistent tones; (d) Spectrum of signal in (c).

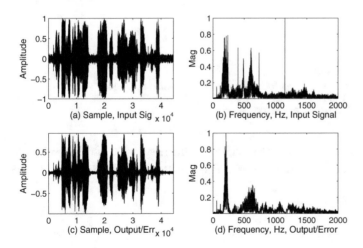

Figure 2.27: (a) Test speech signal having persistent tones added at frequencies of 440 Hz, 740 Hz, and 1150 Hz; (b) Spectrum of test signal at (a); (c) Output/Error signal, i.e., the test signal after being filtered by a Dual-H LMS system using a decorrelating delay designed to assist in removing persistent tones; (d) Spectrum of Output/Error signal. Note the much improved attenuation of the persistent tones compared to that of the previous figure.

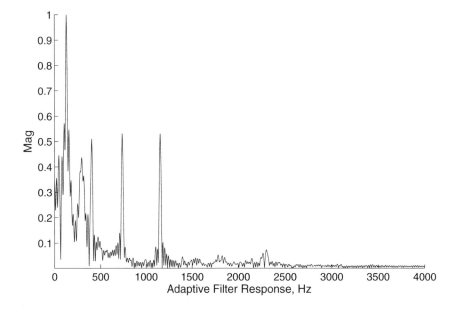

Figure 2.28: Frequency response of the best set of tap weights for the length-400 LMS adaptive filter that is computed to remove simultaneous persistent tones at frequencies of 400 Hz, 740 Hz, and 1150 Hz.

APPENDIX A

Software for Use with this Book

A.1 FILE TYPES AND NAMING CONVENTIONS

The text of this book describes many computer programs or scripts to perform computations that illustrate the various signal processing concepts discussed. The computer language used is usually referred to as **m-code** (or as an **m-file** when in file form, using the file extension **.m**) in MATLAB -related literature or discussions, and as **MathScript** in LabVIEW-related discussions (the terms are used interchangeably in this book).

The MATLAB and LabVIEW implementations of m-code (or MathScript) differ slightly (Lab-VIEW's version, for example, at the time of this writing, does not implement Handle Graphics, as does MATLAB).

The book contains mostly scripts that have been tested to run on both MATLAB and Lab-VIEW; these scripts all begin with the letters **LV** and end with the file extension **.m**. Additionally, scripts starting with the letters **LVx** are intended as exercises, in which the student is guided to write the code (the author's solutions, however, are included in the software package and will run when properly called on the Command Line).

Examples are:

LVPlotUnitImpSeq.m

LVxComplexPowerSeries.m

There are also a small number m-files that will run only in MATLAB, as of this writing. They all begin with the letters *ML*. An example is:

ML_SinglePole.m

Additionally, there are a number of LabVIEW Virtual Instruments (VIs) that demonstrate various concepts or properties of signal processing. These have file names that all begin with the letters *Demo* and end with the file extension *.vi*. An example is:

DemoComplexPowerSeriesVI.vi

Finally, there are several sound files that are used with some of the exercises; these are all in the .wav format. An example is:

drwatsonSR4K.wav

A.2   DOWNLOADING THE SOFTWARE

All of the software files needed for use with the book are available for download from the following website:

http://www.morganclaypool.com/page/isen

The entire software package should be stored in a single folder on the user's computer, and the full file name of the folder must be placed on the MATLAB or LabVIEW search path in accordance with the instructions provided by the respective software vendor.

A.3   USING THE SOFTWARE

In MATLAB, once the folder containing the software has been placed on the search path, any script may be run by typing the name (without the file extension, but with any necessary input arguments in parentheses) on the Command Line in the Command Window and pressing *Return*.

In LabVIEW, from the Getting Started window, select MathScript Window from the Tools menu, and the Command Window will be found in the lower left area of the MathScript window. Enter the script name (without the file extension, but with any necessary input arguments in parentheses) in the Command Window and press *Return*. This procedure is essentially the same as that for MATLAB.

Example calls that can be entered on the Command Line and run are

LVAliasing(100,1002)

LV_FFT(8,0)

In the text, many "live" calls (like those just given) are found. All such calls are in boldface as shown in the examples above. When using an electronic version of the book, these can usually be copied and pasted into the Command Line of MATLAB or LabVIEW and run by pressing *Return*. When using a printed copy of the book, it is possible to manually type function calls into the Command Line, but there is also one stored m-file (in the software download package) per chapter that contains clean copies of all the m-code examples from the text of the respective chapter, suitable for copying (these files are described more completely below in the section entitled "Multi-line m-code examples"). There are two general types of m-code examples, single-line function calls and multi-line code examples. Both are discussed immediately below.

A.4   SINGLE-LINE FUNCTION CALLS

The first type of script mentioned above, a named- or defined-function script, is one in which a function is defined; it starts with the word "function" and includes the following, from left to right:

any output arguments, the equal sign, the function name, and, in parentheses immediately following the function name, any input arguments. The function name must always be identical to the file name. An example of a named-function script, is as follows:

function nY = LVMakePeriodicSeq(y,N)
% LVMakePeriodicSeq([1 2 3 4],2)
y = y(:); nY = y*([ones(1,N)]); nY = nY(:)';

For the above function, the output argument is *nY*, the function name is *LVMakePeriodicSeq*, and there are two input arguments, *y* and *N*, that must be supplied with a call to run the function. Functions, in order to be used, must be stored in file form, i.e., as an m-file. The function *LVMakePeriodicSeq* can have only one corresponding file name, which is

LVMakePeriodicSeq.m

In the code above, note that the function definition is on the first line, and an example call that you can paste into the Command Line (after removing or simply not copying the percent sign at the beginning of the line, which marks the line as a comment line) and run by pressing *Return*. Thus you would enter on the Command Line the following, and then press *Return*:

nY = LVMakePeriodicSeq([1,2,3,4],2)

In the above call, note that the output argument has been included; if you do not want the value (or array of values) for the output variable to be displayed in the Command window, place a semicolon after the call:

nY = LVMakePeriodicSeq([1,2,3,4],2);

If you want to see, for example, just the first five values of the output, use the above code to suppress the entire output, and then call for just the number of values that you want to see in the Command window:

nY = LVMakePeriodicSeq([1,2,3,4],2);nY1to5 = nY(1:5)

The result from making the above call is

nY1to5 = [1,2,3,4,1]

A.5   MULTI-LINE M-CODE EXAMPLES

There are also entire multi-line scripts in the text that appear in boldface type; they may or may not include named-functions, but there is always m-code with them in excess of that needed to make a simple function-call. An example might be

N=54; k = 9; x = cos(2*pi*k*(0:1:N-1)/N);
LVFreqResp(x, 500)

Note in the above that there is a named-function (*LVFreqResp*) call, preceded by m-code to define an input argument for the call. Code segments like that above must either be (completely) copied and pasted into the Command Line or manually typed into the Command Line. Copy-and-Paste can often be successfully done directly from a pdf version of the book. This often results in problems (described below), and accordingly, an m-file containing clean copies of most m-code programs from each chapter is supplied with the software package. Most of the calls or multi-line m-code examples from the text that the reader might wish to make are present in m-files such as

McodeVolume1Chapter4.m

McodeVolume2Chapter3.m

and so forth. There is one such file for each chapter of each book, except Chapter 1 of Volume I, which has no m-code examples.

A.6   HOW TO SUCCESSFULLY COPY-AND-PASTE M-CODE

M-code can usually be copied directly from a pdf copy of the book, although a number of minor, easily correctible problems can occur. Two characters, the symbol for raising a number to a power, the circumflex ˆ, and the symbol for vector or matrix transposition, the apostrophe or single quote mark ', are coded for pdf using characters that are non-native to m-code. While these two symbols may look proper in the pdf file, when pasted into the Command line of MATLAB, they will appear in red.

A first way to avoid this copying problem, of course, is simply to use the m-code files described above to copy m-code from. This is probably the most time-efficient method of handling the problem—avoiding it altogether.

A second method to correct the circumflex-and-single-quote problem, if you do want to copy directly from a pdf document, is to simply replace each offending character (circumflex or single quote) by the equivalent one typed from your keyboard. When proper, all such characters will appear in black rather than red in MATLAB. In LabVIEW, the pre-compiler will throw an error at the first such character and cite the line and column number of its location. Simply manually retype/replace each offending character. Since there are usually no more than a few such characters, manually replacing/retyping is quite fast.

Yet a third way (which is usually more time consuming than the two methods described above) to correct the circumflex and apostrophe is to use the function *Reformat*, which is supplied with the software package. To use it, all the copied code from the pdf file is reformatted by hand into one horizontal line, with delimiters (commas or semicolons) inserted (if not already present) where lines have been concatenated. For example, suppose you had copied

```
n = 0:1:4;
y = 2.^n
stem(n,y);
```

where the circumflex is the improper version for use in m-code. We reformat the code into one horizontal line, adding a comma after the second line (a semicolon suppresses computed output on the Command line, while a comma does not), and enclose this string with apostrophes (or single quotes), as shown, where *Reformat* corrects the improper circumflex and *eval* evaluates the string, i.e., runs the code.

$$\textbf{eval(Reformat('n=0:1:4;y=2.^n;stem(n,y)'))}$$

Occasionally, when copying from the pdf file, essential blank spaces are dropped in the copied result and it is necessary to identify where this has happened and restore the missing space. A common place that this occurs is after a "for" statement. The usual error returned when trying to run the code is that there is an unmatched "end" statement or that there has been an improper use of the reserved word "end". This is caused by the elision of the "for" statement with the ensuing code and is easily corrected by restoring the missing blank space after the "for" statement. Note that the function *Reformat* does not correct for this problem.

A.7   LEARNING TO USE M-CODE

While the intent of this book is to teach the principles of digital signal processing rather than the use of m-code per se, the reader will find that the scripts provided in the text and with the software package will provide many examples of m-code programming starting with simple scripts and functions early in the book to much more involved scripts later in the book, including scripts for use with MATLAB that make extensive use of MATLAB objects such as push buttons, edit boxes, drop-down menus, etc.

Thus the complexity of the m-code examples and exercises progresses throughout the book apace with the complexity of signal processing concepts presented. It is unlikely that the reader or student will find it necessary to separately or explicitly study m-code programming, although it will occasionally be necessary and useful to use the online MATLAB or LabVIEW help files for explanation of the use of, or call syntax of, various built-in functions.

A.8   WHAT YOU NEED WITH MATLAB AND LABVIEW

If you are using a professional edition of MATLAB, you'll need the Signal Processing Toolbox in addition to MATLAB itself. The student version of MATLAB includes the Signal Processing Toolbox.

If you are using either the student or professional edition of LabVIEW, it must be at least Version 8.5 to run the m-files that accompany this book, and to properly run the VIs you'll need the Control Design Toolkit or the newer Control Design and Simulation Module (which is included in the student version of LabVIEW).

APPENDIX B

Vector/Matrix Operations in M-Code

B.1 ROW AND COLUMN VECTORS

Vectors may be either row vectors or column vectors. A typical row vector in m-code might be [3 -1 2 4] or [3,-1,2, 4] (elements in a row can be separated by either commas or spaces), and would appear conventionally as a row:

$$\begin{bmatrix} 3 & -1 & 2 & 4 \end{bmatrix}$$

The same, notated as a column vector, would be [3,-1,2,4]' or [3; -1; 2; 4], where the semicolon sets off different matrix rows:

$$\begin{bmatrix} 3 \\ -1 \\ 2 \\ 4 \end{bmatrix}$$

Notated on paper, a row vector has one row and plural columns, whereas a column vector appears as one column with plural rows.

B.2 VECTOR PRODUCTS

B.2.1 INNER PRODUCT

A row vector and a column vector of the same length as the row vector can be multiplied two different ways, to yield two different results. With the row vector on the left and the column vector on the right,

$$\begin{bmatrix} 1 & 2 & 3 & 4 \end{bmatrix} \begin{bmatrix} 4 \\ 3 \\ 2 \\ 1 \end{bmatrix} = 20$$

corresponding elements of each vector are multiplied, and all products are summed. This is called the **Inner Product**. A typical computation would be

$$[1, 2, 3, 4] * [4; 3; 2; 1] = (1)(4) + (2)(3) + (3)(2) + (4)(1) = 20$$

B.2.2   OUTER PRODUCT

An **Outer Product** results from placing the column vector on the left, and the row vector on the right:

$$\begin{bmatrix} 4 \\ 3 \\ 2 \\ 1 \end{bmatrix} \begin{bmatrix} 1 & 2 & 3 & 4 \end{bmatrix} = \begin{bmatrix} 4 & 8 & 12 & 16 \\ 3 & 6 & 9 & 12 \\ 2 & 4 & 6 & 8 \\ 1 & 2 & 3 & 4 \end{bmatrix}$$

The computation is as follows:

$$[4; 3; 2; 1] * [1, 2, 3, 4] = [4, 3, 2, 1; 8, 6, 4, 2; 12, 9, 6, 3; 16, 12, 8, 4]$$

Note that each column in the output matrix is the column of the input column vector, scaled by a column (which is a single value) in the row vector.

B.2.3   PRODUCT OF CORRESPONDING VALUES

Two vectors (or matrices) of exactly the same dimensions may be multiplied on a value-by-value basis by using the notation " .* " (a period followed by an asterisk). Thus two row vectors or two column vectors can be multiplied in this way, and result in a row vector or column vector having the same length as the original two vectors. For example, for two column vectors, we get

$$[1; 2; 3]. * [4; 5; 6] = [4; 10; 18]$$

and for row vectors, we get

$$[1, 2, 3]. * [4, 5, 6] = [4, 10, 18]$$

B.3   MATRIX MULTIPLIED BY A VECTOR OR MATRIX

An m by n matrix, meaning a matrix having m rows and n columns, can be multiplied from the right by an n by 1 column vector, which results in an m by 1 column vector. For example,

$$[1, 2, 1; 2, 1, 2] * [4; 5; 6] = [20; 25]$$

Or, written in standard matrix form:

$$\begin{bmatrix} 1 & 2 & 1 \\ 2 & 1 & 2 \end{bmatrix} \begin{bmatrix} 4 \\ 5 \\ 6 \end{bmatrix} = \begin{bmatrix} 4 \\ 8 \end{bmatrix} + \begin{bmatrix} 10 \\ 5 \end{bmatrix} + \begin{bmatrix} 6 \\ 12 \end{bmatrix} = \begin{bmatrix} 20 \\ 25 \end{bmatrix} \tag{B.1}$$

An m by n matrix can be multiplied from the right by an n by p matrix, resulting in an m by p matrix. Each column of the n by p matrix operates on the m by n matrix as shown in (B.1), and creates another column in the n by p output matrix.

B.4   MATRIX INVERSE AND PSEUDO-INVERSE

Consider the matrix equation

$$\begin{bmatrix} 1 & 4 \\ 3 & -2 \end{bmatrix} \begin{bmatrix} a \\ b \end{bmatrix} = \begin{bmatrix} -2 \\ 3 \end{bmatrix} \tag{B.2}$$

which can be symbolically represented as

$$[M][V] = [C]$$

or simply

$$MV = C$$

and which represents the system of two equations

$$a + 4b = -2$$

$$3a - 2b = 3$$

that can be solved, for example, by scaling the upper equation by -3 and adding to the lower equation

$$-3a - 12b = 6$$

$$3a - 2b = 3$$

which yields

$$-14b = 9$$

or

$$b = -9/14$$

and

$$a = 4/7$$

The inverse of a matrix M is defined as M^{-1} such that

$$MM^{-1} = I$$

where I is called the Identity matrix and consists of all zeros except for the left-to-right downsloping diagonal which is all ones. The Identity matrix is so-called since, for example,

$$\begin{bmatrix} 1 & 0 \\ 0 & 1 \end{bmatrix} \begin{bmatrix} a \\ b \end{bmatrix} = \begin{bmatrix} a \\ b \end{bmatrix}$$

The pseudo-inverse M^{-1} of a matrix M is defined such that

$$M^{-1}M = I$$

System B.2 can also be solved by use of the pseudo-inverse

$$\left[M^{-1} \right] [M][V] = \left[M^{-1} \right] [C]$$

which yields

$$[I][V] = V = \left[M^{-1} \right] [C]$$

In concrete terms, we get

$$\left[M^{-1} \right] \begin{bmatrix} 1 & 4 \\ 3 & -2 \end{bmatrix} \begin{bmatrix} a \\ b \end{bmatrix} = \left[M^{-1} \right] \begin{bmatrix} -2 \\ 3 \end{bmatrix} \tag{B.3}$$

which reduces to

$$\begin{bmatrix} a \\ b \end{bmatrix} = \left[M^{-1} \right] \begin{bmatrix} -2 \\ 3 \end{bmatrix}$$

We can compute the pseudo-inverse M^{-1} and the final solution using the built-in MathScript function $pinv$:

```
M = [1,4;3,-2];
P = pinv(M)
ans = P*[-2;3]
```

which yields

$$P = \begin{bmatrix} 0.1429 & 0.2857 \\ 0.2143 & -0.0714 \end{bmatrix}$$

and therefore

$$\begin{bmatrix} a \\ b \end{bmatrix} = \begin{bmatrix} 0.1429 & 0.2857 \\ 0.2143 & -0.0714 \end{bmatrix} \begin{bmatrix} -2 \\ 3 \end{bmatrix}$$

which yields $a = 0.5714$ and $b = -0.6429$ which are the same as 4/7 and -9/14, respectively. A unique solution is possible only when M is square and all rows linearly independent.(a linearly independent row cannot be formed or does not consist solely of a linear combination of other rows in the matrix).

Biography

Forester W. Isen received the B.S. degree from the U. S. Naval Academy in 1971 (majoring in mathematics with additional studies in physics and engineering), and the M. Eng. (EE) degree from the University of Louisville in 1978, and spent a career dealing with intellectual property matters at a government agency working in, and then supervising, the examination and consideration of both technical and legal matters pertaining to the granting of patent rights in the areas of electronic music, horology, and audio and telephony systems (AM and FM stereo, hearing aids, transducer structures, Active Noise Cancellation, PA Systems, Equalizers, Echo Cancellers, etc.). Since retiring from government service at the end of 2004, he worked during 2005 as a consultant in database development, and then subsequently spent several years writing the four-volume series DSP for MATLAB and LabVIEW, calling on his many years of practical experience to create a book on DSP fundamentals that includes not only traditional mathematics and exercises, but "first principle" views and explanations that promote the reader's understanding of the material from an intuitive and practical point of view, as well as a large number of accompanying scripts (to be run on MATLAB or LabVIEW) designed to bring to life the many signal processing concepts discussed in the series.